COVID-19:

FROM CHAOS TO CURE

The biology behind the fight against the novel coronavirus

DALE YUZUKI, M.A., M.ED.

COVID-19: From Chaos to Cure

The biology behind the fight against the novel coronavirus

ISBN 978-2-5445-1225-2 *Hardcover*

ISBN 978-2-5445-1225-6 *Paperback*

ISBN 978-2-5445-1225-8 *Ebook*

Edited by Jennifer Collins and India Hammond

Interior layout by Michael Boalch

Cover by aksaramantra of 99designs

To my dear wife Momo,

Without whom much in my life would not have been possible.

"If you're going through hell, keep going."

—Winston Churchill

CONTENTS

TABLE OF FIGURES

INTRODUCTION TO COVID-19: FROM CHAOS TO CURE

There over 44 million cases and 1.1 million deaths worldwide as of October 2020. Over 230,000 deaths in the United States make the nightly news. By the time you have read these words, these numbers will undoubtedly have climbed higher.

News from countries facing possible "second waves" stir fear, social-media updates amplify confusion, and politician's bold promises are greeted with suspicion. As a non-scientist, how are you supposed to make sense of all of this?

Have you wished you had someone with a scientific background who could help break down some of the basics regarding how the COVID-19 pandemic is being addressed? Have you wished you had someone who would not comment endlessly on the politics of policies and governance, but just give you the information you needed? Have you wished you had some additional background information to calm you through the panic and fear-mongering happening around you?

You do not have a background in biology; you took a class or two in high school or college, but it seems so far away from the terms being thrown around in the articles you read and the news you watch. Then, as you are watching the news, there is a segment describing something called a PCR test, with technicians in front of trays of little plastic vials and a piece of rectangular equipment that has plastic tubing sprouting from it. A doctor is discussing the latest experimental treatment; its name is barely pronounceable and the treatment process is not even remotely comprehensible. A vaccine is being developed by a company nobody has ever heard of before.

I am here to tell you that you don't need to argue in circles, and you don't need to be led astray by anecdotes; instead, you can sift through the scientific breakthroughs and keep an eye out for genuine advances, keeping you calm even through a pandemic.

That is what this book is for.

To become aware of how science is tackling this worldwide crisis, we must start from the beginning. Although we won't go through the details of Mendel's discoveries, we will begin with the structure of DNA and give a primer on how scientists' knowledge of DNA grew and expanded, along with discussing the tools that allowed scientists to manipulate and engineer DNA in sophisticated ways.

In addition to this, we will follow an outline of related Nobel Prizes as a convenient shorthand for the major discoveries of biology and applied biotechnology: the proposal of DNA structure, the invention of determining DNA sequence, the discovery of enzymes that cut DNA at particular sequences, the invention of PCR, the discovery of antibodies, and the invention of being able to produce engineered antibodies.

Along the way, we will also delve into the mysteries of the immune system and the history of vaccines. Throughout these discussions, the biology and biotechnology will be related back to the coronavirus' transmission, detection, and COVID-19 itself.

This book is not meant to help diagnose, treat, or prevent disease, it is designed to help you to understand COVID-19 diagnostics, COVID-19 treatments, and the current worldwide effort underway to produce a vaccine that will bring the end of this collective nightmare. With this understanding, you will be able to put new advances and information into the appropriate context.

It will take some patience on your part. In the first section of the book, we will cover some 70 years of breakthroughs and the development of industrial biotechnology from its earliest discoveries on to its most current iteration.

As one who has first-hand experience with the work required to launch a diagnostic test, who has sold scientific equipment and chemicals to pharmaceutical research scientists, and as an advisor to a DNA-based vaccine start-up, I combine over 25 years in the research and commercial worlds of science with a background in teaching, and I will be your guide.

We are currently observing applied biotechnology literally saving the world from death. You have an opportunity now to learn about how this is being accomplished.

CHAPTER 1: THE HUMILITY OF SCIENCE

Most human beings have an almost infinite capacity for taking things for granted.

— Aldous Huxley

I knew that going back to graduate school after a few years of full-time work was going to be difficult, but I hadn't thought it would be this tough. I faced the pressures of time, of opportunities lost, of finances, and of living like a hermit, not to mention the nature of a competitive program where I wanted to show myself as worthy of that degree.

It was another late night in the laboratory for me. It was 1989, and I was heading back downstairs into a well-lit basement with benches crowded with devices (power supplies, acrylic boxes of various sizes, and sleeves of clear plastic Petri dishes). No one else was there at 2:30 a.m.

I had bicycled back to the lab earlier in the evening, pulled a few plates from the fridge labeled Amp⁺ which I had prepared the day before, lit the Bunsen burner, and dipped the bent glass rod into a dish of alcohol to sterilize it. Spinning the plastic plate on a rotating device, tube after tube of transformed *E. coli* spread out, and then they went off to the 37°C incubator.

Bicycling the few miles back home in the late Spring San Francisco air was exhilarating. Way past midnight, I felt as though I was accelerating time as I figured out the next steps in my newest cloning project — imagining the restriction enzyme digest, separating the target DNA fragment on an agarose gel, purifying it afterwards, and then setting up the ligation reaction.

The following morning was bright and sunny. After four hours of sleep, I got up to teach an 8:00 a.m. undergraduate biology lab. Not a lot of prep work was necessary, as this session was the equivalent of teaching out of a cookbook, and the students were a lot of fun.

One older student had returned to school to work on additional undergraduate requirements. It was clear he was going places — he had a successful career in software development and had returned to San Francisco State to go to medical school. He was at SFSU for a few years to take the prerequisite biology, inorganic chemistry, biochemistry, and organic chemistry coursework, followed by the MCAT (Medical College Admissions Test).

Charlie was fun to teach — especially for me, a former high school Honors Chemistry and Biology teacher. The reason he was fun was his incisive questioning; at the college level, the edges and frontiers of science could be explored, investigated, and pushed a little further.

At the end of this class, Charlie asked, "How's the research going?"

"I'm getting to the exciting part," I replied. "If the ligation worked, it means I've got the fragment I've been trying to get at for the past six months."

A ligation is where fragments of DNA are shuffled into the appropriate position and glued together, and I had been having difficulty getting this step to work the way it was supposed to. Calls into the company that sold the enzyme ligase had given me a few suggestions, and a few lab-mates had also given me their ideas on why it was not working. At that point, I had accumulated plenty of days where the ampicillin plates (those clear plastic dishes with transparent yellow agar growth medium) would be empty the following day. Another day's work down the drain.

Ampicillin, an antibiotic, disallows any bacteria that does not have the engineered plasmid in it. There are places engineered into these microscopic loops of DNA where you can cut and paste, just like in any computer program. The ligase performs the final step in pasting the connecting pieces of DNA together. These plasmids have an ampicillin resistance gene and a properly working ligation reaction would confer that resistance to the specially designed and specifically treated bacteria I had plated onto the Amp$^+$ plates only 6 hours before.

The moment of truth came right after the undergraduate lab session.

Opening up the 37°C incubator, I eagerly picked up the stack of ampicillin plates.

The plates were littered with dozens of blue and white dots, which were around a millimeter or two across, each about the size of a pinhead. I just suppressed yelling aloud. There were not many days working in the lab where you felt elated, so I treasured this one.

Any bacteria that survived on the Amp$^+$ plate had the plasmid with the ampicillin resistance gene in it. Any bacteria that had turned blue showed that that particular colony (called a clone, this is a single bacteria that has multiplied many million-fold) contained a plasmid without any gene interrupting another gene called beta-galactosidase. In normal plasmids, the beta-galactosidase gene codes for a protein that metabolizes a dye called X-gal to turn blue. Blue colonies were not good, as they did not have the stretch of DNA that I wanted to capture. The white colonies were the good ones; the bacteria contained plasmids where the beta-galactosidase gene had been disrupted by the engineered piece inside it.

I would choose a few of the white ones for further analysis, scooping up the small dot of bacteria with a sterile wire loop, then stirring it in a test tube with around a half-inch of nutrient broth.

In one of those tubes was a group of bacteria containing what I was after: a snippet of DNA that would become a Master's Thesis. One technique after another, I had learned how to cut DNA, how to stitch it together, how to sequence the DNA, and how to amplify it up from minuscule picograms.

It is these technologies that are fully brought to bear when tackling the problem of a pandemic: how to detect infected individuals, how to develop therapies and perhaps even cures for severe COVID-19, and how to bring a vaccine to market in record time.

DNA TECHNOLOGIES ON FULL DISPLAY IN THE AGE OF COVID-19

At its essence, molecular biology is being able to manipulate DNA for a desired effect, and in this pandemic, all DNA technology is on full display. Detecting live virus is far too complicated and dangerous; therefore, sensitive DNA technologies such as the polymerase chain reaction (PCR) are utilized for detection of the SARS-CoV-2 nucleic acid. Antibodies are sequenced, characterized, and engineered as promising therapeutics to both treat and potentially prevent infection. No less than five distinct technical approaches are currently being utilized in a global effort, with over 200 vaccines and vaccine candidates in various stages of development.

We will take a closer look at these fundamental building blocks and how they directly relate to the diagnostics technology, the therapeutic development, and the vaccine development against COVID-19 in the coming chapters. But first, we must take a look at how science in general progresses.

SCIENCE IS HARD...AND METHODICAL

Science is hard and, at its essence, science is humble. Hard-fought knowledge gained at the edges of scientific understanding will only extend knowledge in a tiny area. Fundamental breakthroughs and new realms of knowledge open up infrequently. Sometimes, a new opening will unfurl into a new geography, and other times, the discovery of new knowledge is only the window to a small, unique archipelago. All too often, it's just a tiny little sandbar — an observation that doesn't lead anywhere.

In science, there is something called "The Streetlight Effect."

Figure 1: The Streetlight Effect

A person (typically a drunkard) is looking for their wallet under a lamp post. A passer-by asks, "Have you looked in any other places?" To which the person replies, "No, I'm looking here because the light is better."

Scientists are world-class experts within the scope of their narrow area of expertise, working at the edges of their specialty to increase what is known. Steve Jobs once said that he wanted to make a 'dent in the universe' — and in that spirit, scientists make hard-won progress every day[1]. Some dents are bigger than others and sometimes small movements lead to much larger effects.

You may be familiar with the scientific method of a three-part cardboard poster from a grade school science fair, either as one who made one or as a parent of a child tasked with making one. The origin of the method comes from the Arab world of the 12ᵗʰ century, from a polymath named Ibn al-Haytham (Latinized name: Alhazen) who wrote,

> *A person who studies scientific books with a view of knowing the real facts ought to turn himself into an opponent of everything he studies; he should thoroughly assess its main as well as its margin parts, and oppose it from every point of view and all its aspects…If he takes this course, the real facts will be revealed to him.*

— **Shukūk ʿalā Baṭlamyūs ("Doubts About Ptolemy")**

While this may seem counterintuitive, you can see the pride of knowledge displayed in the current conflicting viewpoints of experts, public leaders, and informal discourse as we work through the coronavirus pandemic. Instead of making dents, we have witnessed the birth of a new, chattering class of instant experts bloviating on mass media. With social-media amplification, narrow-mindedness and biases are the norms. Ask yourself, when was the last time you heard someone say 'I don't know' in a televised interview or video clip?

al-Haytham refers to "real facts." At its heart, the scientific method finds its way through the darkness by expanding a circle of knowledge through hypotheses and testing, repeating the process over and over again with additional facts, data, and observation. This method is humble at its core. A quote familiar to many of you could speak directly to the process:

> *But he that is greatest among you shall be your servant. And whosoever shall exalt himself shall be abased; and he that shall humble himself shall be exalted.*

— **Matthew 23:11-12**

While you will not find this humility in public discourse, it remains a fundamental attitude within the realm of science. A hypothesis is used to fit the data, in all of its incomplete forms, until a collection of hypotheses can be amalgamated into a model. Alternatively, conflicting data will change such a hypothesis in mid-stream, eventually forming an enlarged model. This model could be used to make predictions and/or changed as new data emerges. Often enough, a completely new model must be invented to account for what has been learned.

THE WORLDWIDE PANDEMIC MEETS SCIENCE

In this Severe Acute Respiratory Syndrome CoronaVirus 2 (SARS-CoV-2) pandemic, many edges of science are being expanded at the same time. Scientists are using their current models of complex systems and applying them to a real-world problem where there are millions of lives, trillions of dollars, and untold suffering at stake.

In broad strokes, you can look at the pandemic as an emergency, such as a brush fire in a large area traveling rapidly from home to home. Diagnostics — whether we speak of a chest X-ray with the tell-tale 'ground-glass' appearance or a nasal-cavity swab from a drive-through test site — amount to the fire alarm. This is the warning system telling us to take action.

The calling in of firefighters and their equipment equates to therapeutics, of which at present, only a handful exist — although more are on the way.

Vaccines are the final answer, a protective layer. In the brush fire scenario, imagine five days of constant rain preventing any new fires from starting.

As far as diagnostics go, the new SARS-CoV-2 tests are based on tried-and-true technologies. For example, think of the at-home plastic test strip pregnancy test with the "–" and "+" readouts. This is a Lateral Flow ImmunoAssay (LFIA), which relies on looking for the telltale signs of a particular hormone in the urine that comes from a growing fetus (this hormone called human chorionic gonadotropin or HCG for short). The change in color detects the presence of HCG with a specific antibody preloaded on the cassette. It costs less than a dollar to manufacture!

Today, molecular diagnostic testing for SARS-CoV-2 requires a much more complicated methodology and equipment (which I promise we will cover the details of in a later section). Yet innovation using molecular methods (either detecting the viral RNA or viral antigen) combined with convenient and cheap at-home methods is well underway by a number of companies that are using a variety of molecular tricks.

This method of pushing the edges of the known in new combinations is a common theme. Until now, we have not had the urgency of public health and economic emergencies driving us to move as quickly as possible. But the processes here were already underway before this pandemic, with other public health challenges in mind; the combination of molecular methods with the LFIA cassette has been worked on by several groups over the past three years or so. The pandemic has simply accelerated all of the timelines.

The same theme of extending the edges and boundaries of science is observed in the development of therapeutics and vaccines. With underlying knowledge and the infrastructure of monoclonal antibodies being put to use with urgency, new combinations of ideas are being tested and refined. The course of developing novel vaccines (there are five generally-used approaches, all of which are being tried at once with over 165 in pre-clinical development and 36 in Phase I to Phase III clinical trials as of the Fall of 2020) is being accelerated in creative ways never tried before.

A LITTLE HISTORY

Attitude is everything.

—**Anonymous**

In the second century AD, Galen of Pergamon made observations of animals (mainly monkeys and pigs) and formulated theories on the circulatory, nervous, and muscle systems. Galen's ideas then stood for over a thousand years. Even today, terms such as "sanguine" come from Galen's belief that human moods are caused by an imbalance between four bodily humours. He carefully observed humans in their diseased state as a physician and conducted experiments and observations on animals; he thus dominated Western medical science for 1,300 years.

In the Renaissance, it was Michelangelo who snuck into a morgue to examine underlying anatomy. He acknowledged the existing limits of depicting the human form and sought a more realistic art, carefully observing and drawing the corpses. Ignorance of human anatomy, a stumbling block to progress in art, was slowly melting away.

A few hundred years later, Charles Darwin's five-year voyage to the west coast of South America on the *HMS Beagle* revealed to him thousands of specimens that he would study for several decades. Observation coupled with careful thought led to the questions that Darwin asked: *Why are there so many kinds of finches in the Galapagos? Where did they come from? What do they have in common? In what important ways do they differ? How could this change have come about?*

These advances in science, whether coming from the ancient Greeks, the Renaissance or the 19th and 20th centuries, were fueled by our innate curiosity and a distinct awareness of our own ignorance.

> *There are known knowns. There are things we know we know. We also know there are known unknowns. That is to say, we know there are some things we do not know. But there are also unknown unknowns, the ones we don't know we don't know.*
>
> **— Secretary of Defense Donald Rumsfeld, 2002 Department of Defense News Briefing**[2]

The literal translation of the word "science" is "to know" in Latin. The history of science is a history of mankind trying to reveal the known unknowns and on the way to such a revelation discovering unknown unknowns. In the translation of space between the two, there is no other attitude that can be taken than that of a student, a learner…one who does not know.

Science starts and continues in the state of unknowing. As a seeker and a learner, try to be skeptical as well as open, unsure, and humble. It is a large request, I know, and not easy to accomplish.

In the past ten years, knowledge of human origins and migrations has exploded, thanks to genomic analysis of ancient DNA coupled with archeological discovery. Neanderthals (*Homo sapiens Neanderthalis*) are known to have co-existed with Homo Sapiens (*Homo sapiens Sapiens*) for almost 20,000 years (47K to 65K years ago) and hence bred human-Neanderthal hybrids. Within the Denisova Cave in the Siberian mountains, the discovery of a single finger bone fragment led to the discovery of a third species of humans that co-existed with Neanderthals and Homo Sapiens for about 10,000 years (44,000 to 54,000 years ago). There is also additional evidence of "ghost" human-like species in our early history.

Where did these other ancient humans come from? Why did they become extinct? What were they like? These are all examples of unknown unknowns becoming known unknowns.

Several international missions to Mars were launched in the Summer of 2020. The U.S. mission is set to analyze a 3-billion-year-old delta for signs of life. It will prepare specimens for eventual retrieval to Earth perhaps as early as 2031. This $2.2 billion investment is a clear demonstration of our own curiosity, our awareness of our own unknowing, and the ability and resources to do something about it.

THE SCIENCE OF A FAST-MOVING DISEASE

Scientists, by nature, are methodical and cautious in making bold statements. Yet the urgency of the moment is unleashing a flood of effort, data, and techniques as they dissect how best to get ahead of a disease that is wreaking havoc on the world.

In order to make sense of this, we will next look deeper into the biotechnology of the science.

CHAPTER 2: THE BIOTECHNOLOGY REVOLUTION IN BRIEF

At lunch Francis [Crick] winged into the Eagle to tell everyone within hearing distance that we had found the secret of life.

— James Watson, in The Double Helix: A Personal Account of the Discovery of the Structure of DNA (1968)

In 1869, Swiss physician Friedrich Miescher observed the pus of wounded soldiers and discovered a large molecule in the nucleus of the cell which he called "nuclein" — later studying the same molecule in salmon sperm.

For most of the following century, biologists would consider proteins with their high level of complexity, variety, size, and chemical properties to be the transmitters of inherited genetic information.

Scientific inquiry moves at its own pace, with its focus on the edges and clues accumulating in a piecemeal fashion.

There are times where the edges of science expand quickly as findings accelerate and the communication of discovery excites the imagination of others to apply similar or new methods for further revelation. The latter half of the 20th century was one of these times.

In 1944 at Rockefeller University, Oswald Avery demonstrated that DNA was the chemical behind the "transforming principle" that could turn non-infective, bacteria-causing pneumonia into an infective strain. It was DNA, not protein, that conferred this key trait for the simple, single-cell organism.

Other scientists quickly pursued various lines of inquiry related to the chemical nature of DNA, its component parts, and, most importantly, how genetic information could be encoded. The Watson and Crick Proposal was built upon a flurry of activity in the nine years ensuing from Avery's observation in 1944, including a key insight from Erwin Chargaff's laboratory at Columbia University: Regardless of the organism in question, the number of adenine

(A) and thymine (T) bases were the same, and the number of guanine (G) and cytosine (C) bases were the same.

James Watson's and Francis Crick's 1953 paper in the journal *Nature* titled "Molecular Structure of Nucleic Acids: Structure for Deoxyribose Nucleic Acid" was modest in its proposal.[3] The single figure in the 1953 article states simply: "This figure is purely diagrammatic." The tentativeness of the author's conclusion is characteristic of scientists: "It has not escaped our notice that the specific pairing we have postulated immediately suggests a possible copying mechanism for the genetic material."

THE KNOWN UNKNOWN

This known unknown — *What is the structure of DNA?* — attracted others to enlarge the field through refining their methods. Creative and observant individuals would modify existing equipment, build new devices, or try a technique in a completely new field in order to answer the question at hand.

The edge of knowledge on the frontier often leads to a peninsula or a barren shore. Other times, it leads to an entirely new world.

This discovery was built upon several preceding lines of investigation, including the X-ray photographs of Rosalind Franklin and Maurice Wilkins (King's College London) and was not immediately celebrated. In hindsight, it led to a veritable explosion in fundamental research, which then turned into applied commercial uses that empowered transformations of industries such as agriculture, drug development, consumer products, and diagnostics.

The structure of DNA was the linchpin for the biotechnology revolution.

Its discovery kicked off an especially fertile time for fundamental developments related to the rearrangement, copying, and other manipulation of DNA, which led to manifold changes in approaches to a number of industries. Of greatest consequence was an array of novel medicines from this kind of capability.

Instead of chemicals (which the pharmaceutical industry calls "small molecules"), biological entities can be used as therapeutic agents, whether they are synthetic hormones, antibodies, gene-therapy-enabling engineered viruses, or, most recently, engineered cells. Newer methods for therapeutic applications continue to be invented, several of which will ultimately be approved as therapies for a wide variety of human diseases.

One modest example of how this potential has transformed consumer products is that it is now possible to modify a dog's diet with specialized food matched to what his or her genes reveal about their particular metabolism.

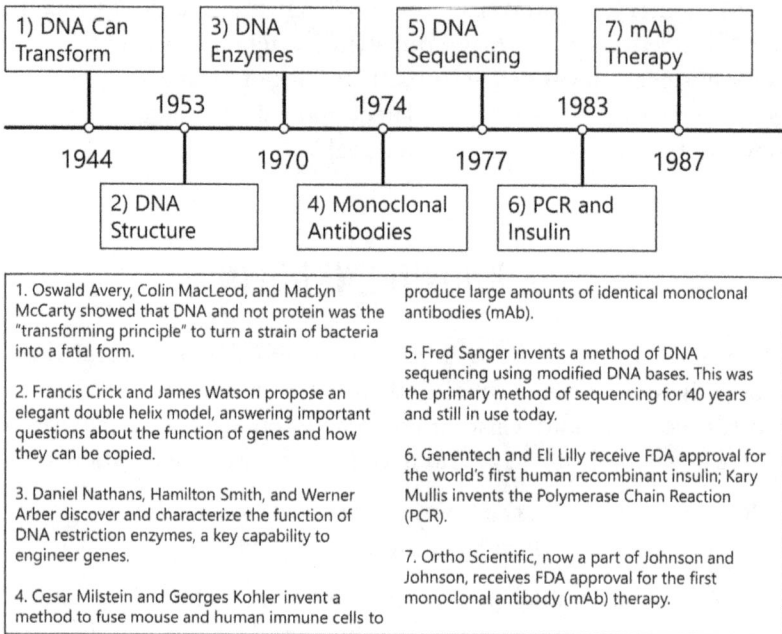

1) DNA Can Transform	3) DNA Enzymes	5) DNA Sequencing	7) mAb Therapy

1953 · 1974 · 1983

1944 · 1970 · 1977 · 1987

2) DNA Structure	4) Monoclonal Antibodies	6) PCR and Insulin

1. Oswald Avery, Colin MacLeod, and Maclyn McCarty showed that DNA and not protein was the "transforming principle" to turn a strain of bacteria into a fatal form.

2. Francis Crick and James Watson propose an elegant double helix model, answering important questions about the function of genes and how they can be copied.

3. Daniel Nathans, Hamilton Smith, and Werner Arber discover and characterize the function of DNA restriction enzymes, a key capability to engineer genes.

4. Cesar Milstein and Georges Kohler invent a method to fuse mouse and human immune cells to produce large amounts of identical monoclonal antibodies (mAb).

5. Fred Sanger invents a method of DNA sequencing using modified DNA bases. This was the primary method of sequencing for 40 years and still in use today.

6. Genentech and Eli Lilly receive FDA approval for the world's first human recombinant insulin; Kary Mullis invents the Polymerase Chain Reaction (PCR).

7. Ortho Scientific, now a part of Johnson and Johnson, receives FDA approval for the first monoclonal antibody (mAb) therapy.

Figure 2: Biotechnology milestones often recognized by a Nobel Prize

In order to understand some of the finer points along the timeline of this explosion of discovery from the 1950s through the 1980s, we will look at several important methods that were then either invented or revealed as a naturally occurring process that could be put to good use.

FOUR KEY ADVANCES

The four key advances to take note of are as follows:

1. The invention of a method to sequence DNA.

2. The discovery of the restriction enzyme.

3. The invention of the Polymerase Chain Reaction (PCR).

4. The discovery of the structure, function, and genetics of antibodies — and how they can be manufactured.

These four discoveries underpin and encapsulate the biotechnology revolution, and they happen to have garnered at least one if not several Nobel Prizes.

In order to understand some of the finer points of the diagnostics, therapeutics, and vaccines being developed to combat SARS-CoV-2 infection and COVID-19 disease, it is important to have a general grasp of what DNA sequencing, restriction enzymes, antibodies, and PCR are, how they work, and how they have enabled modern-day biotechnology. Without any of these four seminal discoveries and inventions (and all of them exhibit a combination of discovery and invention), we would not be where we are today.

I. HOW DNA IS SEQUENCED

The first key underpinning to biotechnology is the ability to sequence DNA. This is the verb "to sequence" — as in "to determine the sequence of DNA" — rather than the noun form of "a sequence of DNA, letters G-A-T-C."

A key early figure to note is Fred Sanger, who published breakthrough work in 1955 that deciphered the 51-amino-acid sequence of the insulin protein. Sanger won the Nobel Prize in Chemistry in 1958 for this achievement.

In 1962, Sanger used radioactive methods to attempt the sequencing of amino-acyl transfer RNAs, which were known to be 75-base-long polymers. This led to a series of inventive and creative experiments, and in 1977, he published a description of a method that is still a mainstay in sequencing more than 40 years later. In what is called di-deoxy chain termination, growing fragments of synthesized DNA will stop at a given base (one vial being all G residues, a second vial being all A residues, and so on and so forth for T and C bases). Each vial's fragments are labeled with radioactivity and separated via thin polyacrylamide gels, and an X-ray film will show an arrangement of radioactive bands corresponding to the DNA sequence.

A word about these "gels," they are a gelatinous mix of chemicals used to separate strands of DNA via an electric current, which are then detected with an X-ray film (if radioactively labeled) or special dyes. As practiced today, the key steps are automated. Each gel is an ultra-thin, hair-like capillary that has been pre-filled with the gel components, and fluorescent-labeled

DNA fragments are identified by a laser coupled to an optical sensor and photomultiplier tube, as they are pulled past the detector by high-voltage DC electricity. A computer labels each DNA base, which is recorded in an electronic file. While today's machinery would be unrecognizable to a 1977-vintage scientist, the basic chemistry remains the same today as it was understood then.

One related technical advance during the 1970s and 1980s was the ability to synthesize DNA; short, single-stranded fragments are called primers, and these primers provide the starting point for a DNA sequencing reaction by providing a starting point for a special enzyme (a polymerase) to elongate a DNA strand. DNA produced by a series of test-tube chemical reactions functions the same as natural DNA.

Engineered DNA can be put into bacteria to produce human insulin or to coax mouse cells into manufacturing human antibodies for therapy.

Machine-made DNA has even been used to create the genome of a simple, unicellular organism. In 2010, a group led by Craig Venter synthesized a copy of a genome of one species of bacteria (*Mycoplasma mycoides*) and transplanted it into a closely related species (*Mycoplasma capricolum*), using yeast as an intermediary for the assembly of the 1.1-million base-pair genome. This field, called synthetic biology, aims to do more than engineer genes; it aims to invent entirely new capabilities by using cells as living factories, perhaps for producing energy, creating novel sensors, or even doing numerical calculations.

II. THE DISCOVERY AND USE OF RESTRICTION ENZYMES

The second major underpinning of biotechnology has been the discovery and widespread use of restriction enzymes, a kind of molecular scissors. Discovered in bacteria as a type of elementary, immune defense mechanism against enemy bacterial viruses (called bacteriophages), these enzymes inside bacteria are constantly scanning for a specific sequence of DNA, and upon sensing them (for example, the DNA bases "GAATTC"), the enzyme will make two cuts of the DNA — one on each side of the phosphate backbone. (The phosphate backbone is the "spiral" on the outside of the DNA structure drawing, while the G-A-T-C bases form the "steps" along the twisted ladder.)

These specific cuts could form a "blunt end"— such as with G-A-A | T-T-C. They could also leave a "sticky end" as seen with G | A-A-T-T-C on one strand of DNA, while its opposite strand would have the cut in this way: G-A-A-T-T | C.

To illustrate both strands, it would look something as follows:

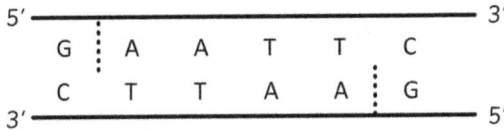

```
5' ─────────────────────────────── 3'
      G ┊ A    A    T    T    C
      C    T    T    A    A ┊ G
   3' ─────────────────────────────── 5'
```

Figure 3: Drawing of a complementary set of DNA strands cut with a restriction enzyme (called Eco RI). The dotted line is where the enzyme cuts the DNA backbone.

The 5' and 3' numbers (read aloud as "five prime" and "three prime") are the way scientists specify the direction of the five-carbon sugar ring (deoxyribose) where phosphate groups are stringing the DNA nucleotides together. As the diagram illustrates, the forward direction of the DNA bases along the top strand of six bases is from left to right (5' to 3' or "five prime to three prime") while for the bottom strand of six bases, forward (still 5' to 3') runs from right to left (in the opposite direction).

The complementary nature of DNA indicates both structure and function. One way to think of it, is if you have a bronze cast of an object, you can make a mold that lays out every detail. Using that mold, you can make a new bronze object that is a duplicate of the original. In turn, that duplicate can be used to make another mold.

With DNA Adenines (A) only pair with Thymines (T), and Guanines (G) only pair with Cytosines (C); this is what Chargaff discovered and called his "rules." Therefore, one strand can make its opposite copy with great accuracy, and that copy, if separate from the original strand, can recreate the original strand through a second set of DNA synthesis.

Now, looking at the restriction enzyme again, we see that the recognition site can cut in a fashion that leaves a jagged edge at a particular location, which looks like this:

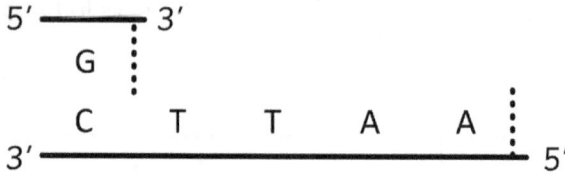

Figure 4: Drawing of a pair of DNA complementary strands with the "sticky end" exposed and ready to combine with a similar end-piece of DNA.

The single bases 3'—T-T-A-A—5' can find another sticky end cut by the same enzyme with the end 5'—A-A-T-T—3' and form a complementary pair. Add an enzyme called a ligase to sew up the nicked phosphate bond backbone and you have just combined two different DNA snippets into one.

These snippets could come from a bacterial plasmid (those small little loops of DNA, ranging from 3,000 bases to invented entities dubbed bacterial artificial chromosomes, which are millions of bases in length), from a plant, from a human gene, or from a virus. They could even be a synthetic stretch of DNA ordered online from a vendor.

By obtaining the DNA sequence of a gene — insulin, for example — and using enzymes for cutting and pasting the gene into the snippets of DNA and then into bacteria (just like those little colonies growing on Amp+ plates, which I mentioned in the first chapter), human insulin could be produced instead of us having to purify bovine insulin from cow organs.

Early on, there was great incentive to make recombinant human insulin (by the word "recombinant," I reference the use of the tools of genetic engineering to splice DNA pieces together). Prior to 1983, insulin in the United States was produced from hog and cattle pancreases. About 56 million animals would provide 19 tons of pancreas organs in order for the medical community to obtain enough porcine and bovine insulin to meet the growing demands of diabetes patients.

Through recombining DNA in this way, Genentech was able to produce in bacteria its first product, a fully human insulin hormone. The sequenced gene from a human was cut with a restriction enzyme, then purified and spliced into a plasmid. Before the gene sequence, there was the appropriate bacterial DNA sequence required for the bacteria's cell machinery to produce insulin's two protein chains as proteins of their own. Each protein was purified and then further processed to make a functional insulin hormone (specifically, two sulfur-containing amino acids had to undergo a chemical reduction reaction combining the two protein strands, forming what is called a disulfide bond).

The world of restriction enzymes includes a collection of types categorized by what they do or how they work. One group is classified by its recognition sequence. In the example of Figures 3 and 4, you can see that the *E. Coli* bacterial species has a naturally occurring enzyme called Eco RI (shorthand for "*E.coli* restriction enzyme Type I"), which recognizes the specific six bases 5'—G-A-A-T-T-C—3' and cuts inside that sequence. There are some enzymes that recognize four bases, others that recognize seven bases, and there are those that recognize a string of six bases and skip the next twenty (regardless of their identity) in order to precisely cut right there. Many laboratories mount wall posters from enzyme providers, offering this information so that scientists can readily obtain exactly the enzyme required to do the job they need to accomplish.

Prominent vendors such as New England Biolabs, Thermo Fisher Scientific's Invitrogen division, and Bio-Rad offer hundreds of different restriction enzymes to their scientist customers.

III. THE POLYMERASE CHAIN REACTION

The third key achievement to be discussed here is the polymerase chain reaction (PCR). This is a method of exponential amplification of DNA that is elegant in its approach.

To start an amplification reaction, a "forward primer" (1) and a "reverse primer" (2) must be designed and manufactured.

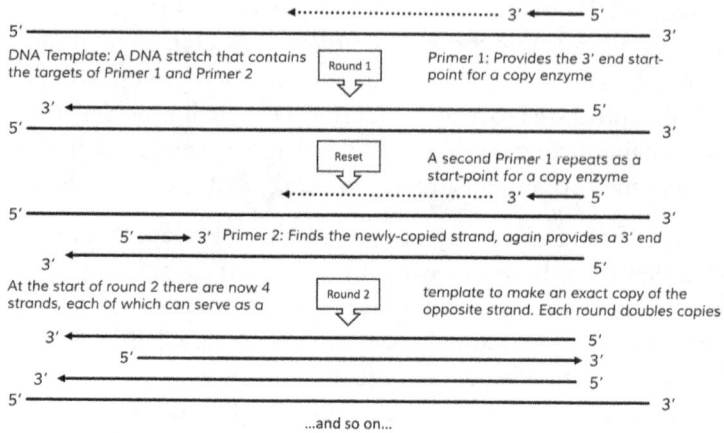

Figure 5: The Polymerase Chain Reaction (PCR) is a process of copying minute amounts of DNA with short synthetic strands of DNA called primers. In each round, these primers find the target sequence and a DNA-copying enzyme manufactures the complementary strand; this strand, in turn, becomes the template for another primer to find.

In round 1, primer 1 extends a complementary ("negative") strand from the 3' end of the primer to the 5' end of the template in the presence of nucleotide building blocks and a DNA polymerase enzyme. The reaction is then heated to near boiling, separating the strands. The complement strand ("negative") binds to primer 2 at the start of round 2 when the reaction mix is cooled. Again in the presence of nucleotide building blocks, this time primer 2 (the "positive" strand) works off the negative strand's 3' end to synthesize a positive strand.

The newly synthesized positive strand is heated again, separating both positive and negative strands, and each can then serve as templates for the complement strand — after cooling and binding to primer 1 and primer 2 respectively again and again.

With each round doubling the number of amplified molecules, PCR can take very few molecules (or even a single molecule) and multiply them to many billions of molecules in a few hours. This process uses specialized heating equipment which is able to heat the reaction mix to 95C in order to melt the strands apart, cool down to a range of 45C to 60C (depending on a number of variables, including the length of the primers and how many G-C bases there are in comparison to A-T ones), and be raised again to 72C.

For those of you wondering, PCR-based testing for the presence of SARS-CoV-2 RNA requires skilled workers with years of experience who can make sure there is no contamination, that the reactions are set up properly, and that the expensive instrumentation is maintained correctly. In addition to this, software to accomplish data analysis must be available to determine Ct values. (Ct values are a measure of the amount of original virus in a sample.)

The lower the Ct number, the higher the amount of virus, which typically lies within the range of 23 to 40 where it exists, where 23 is a very high amount of virus (hundreds of millions of copies) and 40 is a very low amount (in the range of 5 to 10 molecular copies).

We will cover PCR-based testing in an upcoming section, as it is the major method of detection for the presence of the novel coronavirus in the upper respiratory tract.

IV. THE STRUCTURE, GENETICS, AND FUNCTION OF ANTIBODIES

The last capability to be examined in this biotechnology revolution is the ability to understand the structure, underlying genetics, and function of antibodies.

The ability to sequence and synthesize DNA, to cut and paste DNA fragments at will, and to use PCR in manifold ways to further manipulate DNA sequences, is all on display in the process of generating the specific genetic coding for a specific antibody against a specific group of amino acids, and this stretch of specific amino acids are contained within a protein that the body has never seen before. In the case of coronavirus, it would be against the spike protein. (This will be discussed in depth later, along with discussing antibody-based diagnostics, antibody therapeutics, and the body's immune system in response to vaccination.)

Perhaps the most important capability when discussing antibodies is the ability to manufacture identical antibodies (called monoclonal antibodies or mAbs) in large amounts at injection-grade quality, these being designed for therapeutic drugs that work in a fundamentally different way.

Antibodies are one of the true wonders of nature. There are an estimated 10 to the power of 15 (that is, one quadrillion) numbers of different antibodies made for specific molecular targets. The target could be a sugar molecule from a bacteria never found in a human, a fat molecule only found in plants, or a stretch of protein only found in a parasite.

The number of different antibodies is a number so large (that is, on the order of a thousand trillion or a million billion) that it is hard to imagine how many different kinds of antibodies could be created. Indeed, theoretical calculations based solely on the numbers of shuffled gene components raise that quadrillion number (ten to the fifteenth power) another three orders of magnitude to a quintillion (ten to the eighteenth power).

The mechanism by which we obtain such an astounding number of potential antibodies is a matter of gene rearrangement in the white blood cells of the body, specifically the B-cells and T-cells of our immune system. By the numbers, there are a full 51 variable ("V") regions that independently combine in blood cell development to 27 diversity ("D") regions and several different types of constant ("C") regions, as well as 6 joining ("J") segments; this is just for one of the two antibody substructures called the "heavy chain".

The light chain has its own structure of variation (including other numbers of V, D, C, and J regions with their own nuances).

Here, you have two very large numbers of potential combinations of shuffled genes, and these two heavy and light chains combine with each other, adding on yet another layer of complexity. Thus, we have a wonderful and practically infinite number of antibodies.

For any single antibody with a specific target (for example, against the novel coronavirus spike protein), we can produce hybrid cells of both human and mouse origin. These specialized cells, called hybridomas, enable scientists (and biotechnology companies) to produce that single specific antibody on a limitless scale.

Monoclonal antibodies ("mAbs," pronounced "mabs") were first approved for therapy in 1986, but due to the mouse/human hybrid nature of the antibody molecules, they were generally recognized as non-human and failed over time. The human immune system recognized the minor changes between the mouse and human portions of the antibody and generated an immune reaction against it. Fully human antibodies (with no trace of mouse antibody protein present) are now manufactured in large manufacturing volumes using genetically modified mouse cells and recombinant DNA technology.

It goes without saying that much of the biotechnology revolution could not have occurred without the aid of computers. Being able to search a 5,000 base-pair plasmid for a specific EcoRI recognition sequence such as "GAATTC" is automatic for an enzyme, but very hard to do for a human using only their eyes, and what about analyzing a 5,000,000 base-pair bacterial genome for a specific 20 base-pair primer sequence? Or sifting through a 3,200,000,000 base-pair human genome?

Ah, and there it is: the human genome. The crowning achievement opening the door to so much of this Century of Biology. We turn our attention to the human genome next.

CHAPTER 3: THE HUMAN GENOME PROJECT'S BOOST TO BIOTECHNOLOGY

If the 20th century was the century of physics, the 21st century will be the century of biology. While combustion, electricity and nuclear power defined scientific advance in the last century, the new biology of genome research — which will provide the complete genetic blueprint of a species, including the human species — will define the next.

— Craig Venter and Daniel Cohen, "The Century of Biology"[4]

The human genome is organized into 46 chromosomes, where each chromosome is an inherited copy — one called the paternal chromosome originating from your father and the other called the maternal chromosome originating from your mother. Given that there are 3.2 billion bases per haploid genome (DNA letters GATC per maternal or paternal genome, such as those found in an egg or sperm), there are about 6.4 billion bases per individual cell.

Let's do a simple exercise and print out a single 8.5 by 11-inch page of 12-pt Times New Roman letters, printing GATC in random order on the piece of paper. This page amounts to 2,230 characters. If you divide 6.4 billion characters by 2,230 characters yielded per page, this amounts to 2,869 million pages.

A modern home-office monochrome laser printer from a major manufacturer has single-sided print speeds of 30 pages per minute. Those 2,869 million pages would take over 66 days to print with the printer working 24 hours per day, seven days per week.

How much paper does 2,869 million pages represent? A standard 10-ream case of paper has 5,000 sheets, thus 573.8 cases would be needed to print out a single diploid genome. With each case measuring 12 inches by 18 inches by 10 inches (1.25 cubic ft.) this would amount to 717.25 cubic feet, or a stack of paper which is 10 feet by 10 feet and 7 feet high.

Consider lining up all of the pages contiguously starting from Washington, D.C. It would be an enormous task. If you multiply 2,869 million pages by 11 inches, then divide by 12 inches per foot, the result is 2.75 million feet. Dividing by 5,280 feet in a mile equals 520.9 miles. Going 520 miles from Washington, D.C. will get you to Lexington, Kentucky, or Charleston, South Carolina, as well as to Kennebunkport, Maine.

This is a printout of a single genome. Now consider what would happen if you tried to find a single string of letters —"tagggaaggc aaaaacagaa ccaaataaat gtgtgagtca"— along those 2.9 million pages (spaces at every ten bases is a convention to make reading such a string easier), which differs in a single base (so you would look for another string instead —"tagggaaggc aaaaacaaac caaataaatg tgtgagtcat"). If you look at the second group of 10 carefully, you will see a single missing "C" among the first two letters. This missing "C" throws off all of the bases following it and is known as a point deletion; *point* in that it is a single base, not a succession of bases and *deletion* meaning it disappeared rather than another base being substituted in its place. This missing single base (which technically has several categorizations with names like c.2389delG, 2508delG, and p.Glu797fs) is a *BRCA1* mutation shown to be very serious. If a woman was undergoing BRCA genetic testing for inherited cancer risk and the testing showed this single base deletion on both inherited copies, she would have a 30% to 60% risk of developing breast cancer by age 60.[5]

Science communicators from NASA have expressed difficulty in communicating inter-solar distances, and we have a similar problem here. Taking 66 days to fill up half a room's worth of paper, to line up pages from the District of Columbia to Maine — let alone making sense of it or finding a particular 50-character long string — is difficult for our minds to comprehend. Even the cost of printing all these pages is astronomical. At 3 cents per page, a single printed diploid genome would cost $86,000 to print out. One research group has suggested a new term in an article titled "Big Data: Astronomical or Genomical?"[6]

THE HUMAN GENOME PROJECT "MOONSHOT"

The advances described previously, from DNA sequencing to restriction enzymes and PCR, are the methods of discovery: of making hypotheses and of refining models. The human genome was a moonshot over a decade in the making, and since the milestone draft was published in 2001, this massive dataset has unleashed a torrent of additional discovery, new hypotheses, new genetic tests, and refined models.[7] [8]

Instead of exploring our nearest physical extraterrestrial body, retrieving some rocks to study, and leaving behind some equipment and instruments (the last astronaut to visit the moon did so in 1972), the genome moonshot was the landing on a shore of an unexplored continent. Before, we only had an understanding of some of the major landmarks, some models about how some of the mechanics worked, and a few details covering a smattering of locations. Now, with the sequence in hand, we can go deep into unsolved mysteries, such as the genetic underpinnings of cancer or the strong genetic component of mental health disorders like schizophrenia and Alzheimer's disease. We can then use this information to develop novel therapeutics tailored to both the disease and the individual.

The human genome would not have been made possible without simultaneous advances in computational power. Additional mathematical and theoretical underpinnings put that power to use in sifting through literally billions of characters for a unique string of genetic information. 2.8 million pages of data are not helpful unless you can find what you are looking for. A common search algorithm uses an underlying algorithm called the Burroughs Wheeler Transform, which is a method of data transformation first discovered in 1983 and put to use in 1994. This method sped up the computing calculation time needed by over 10-fold.[9]

Additionally, the Human Genome Project could not have made very much progress throughout the 1990s without an industrial component to provide scientists with the enzymes, chemicals, and other reagents and equipment needed to perform the work. Researchers could simply buy a vial of *Eco RI* enzyme from a company called Promega instead of going to the trouble of growing up a batch of the bacterium *Escherichia Coli* and purifying the enzyme themselves.

Throughout the 1990s, as sequencing expanded during the Human Genome Project, there was a race among equipment providers to sequence more DNA at once, as well as sequencing longer stretches of DNA. These providers had an instrument with a proprietary chemical consumables business model, selling an expensive instrument only used with proprietary reagents available from the manufacturer. Companies such as Applied Biosystems and Amersham PLC hired life scientists to become sales and marketing professionals. Journals and trade magazines displayed glossy, full-page advertisements — not only for conferences or a new enzyme, but for new equipment. Software companies offered help in assembling and managing recombinant plasmid collections; chemical companies offered specialized fine chemicals and required buffers; and the centrifuge manufacturer Eppendorf AG not only manufactured centrifuges, but also the small plastic vials that spun around in them.

These same providers today have been struggling with unprecedented demand for their products and services; whether we're talking about PCR instrumentation, where service engineers are required to qualify the instrumentation and ensure it is running correctly; complex bioprocessing plants that grow engineered cells and are churning out monoclonal antibodies for a new COVID-19 therapy; or labs specializing in the synthesis of nanoparticles and RNA for a new approach to vaccine development. Without this infrastructure and industry in place, the pace of development of diagnostics, therapeutics, and vaccines simply would not have been possible.

THE COTTAGE INDUSTRY OF LIFE SCIENCE

In the 1990s, it was not unusual to have a cabinet full of telephone-order catalogs full of enzymes, chemicals, and countless chemically-modified antibodies for different types of detection. Catalogs of equipment were also available: Two were one-stop-shopping tomes from distributors VWR and Fisher Scientific (now part of Avantor and Thermo Fisher Scientific respectively) from which you could order anything, from plasticware for your cell culture experiments, to the laminar flow hood you would need to do your experiment with. You could order the black inert countertop where you would mix up a batch of Tris buffer, along with the high stool you sat on, even the white lab coat embroidered with your name. All required products could come from the same single vendor.

The commercial aspect of biology is a narrow field; it requires a deep understanding of what a scientist's needs are at the bench, what the current market is offering, and the invention, development, or sourcing of the item in question (whether a chemical, an enzyme, a piece of equipment, or a software program) to meet that need. It also requires a way to communicate to that market; in the 1990s, this involved marketing in the form of advertising, trade show appearances, and mentions in the "Methods" sections of influential journal publications. Alongside this messaging to scientists was the local sales representative, whether they came from a gigantic catalog or a specialty antibody supplier, who would pay visits, gauge needs and offer advice, as well as asking customers to try their products.

The toolbox expanded thanks to this commercial incentive for companies to solve problems faced by the life scientist, alongside the computer capability of analyzing increasing amounts of DNA data. Companies' rising R&D budgets grew their abilities to identify unmet needs and invent new methods with new capabilities, often resulting in the industrialization of something that a basic research scientist had invented.

PCR as a fundamental tool in the laboratory spawned an explosion of applications and modifications right from the time of its creation in 1985; as a marker of the value of this invention, the Swiss pharmaceutical company Hoffman-LaRoche, through its subdivision Roche Molecular Systems, paid $300 million for the PCR patent and associated technology from its owner, Cetus Corporation. Other companies manufactured both the equipment and reagents for synthesizing one's own DNA fragments to be used in those PCR machines.

Throughout the late 1980s and the 1990s, there was a rapid expansion in the sale of PCR instrumentation and consumables. There was rocketing demand for enzymes needed for the reaction, as well as the custom synthetic DNA primers required for the specific region to be amplified. The number of inventive techniques for applying the idea of PCR is too many to count. These innovations accompanied plenty of journal articles, in addition to an inchoate internet as a communication tool with email lists (Listservs) and Usenet newsgroups. The Usenet "Methods and Reagents" forum ("BIOSCI/ Bionet methods-reagents") was a useful forum for discussion and advice.

The aforementioned PCR technology found immediate application in paternity and forensics; the ability to positively identify individuals for these applications has since been invaluable. Uses of PCR expanded into applications in archaeology and, perhaps most importantly, diagnostics. While preceding the Human Genome Project, the information PCR provided could quickly be applied to diagnosing genetic disorders and infectious disease — in particular HIV, which overlapped with the discovery of PCR in the 1980s.

The HIV epidemic was a forerunner of the current public health crisis and coincided with the ability of scientists to detect trace amounts of virus using PCR. These same PCR diagnostics are now being used worldwide for the detection of the novel coronavirus.

PCR DIAGNOSTICS AND THE HIV EPIDEMIC

One key advance in the application of PCR to diagnostics was the ability to monitor the polymerase chain reaction while the reaction was ongoing. In the mid-1990s, scientists at Applied Biosystems (now a division of Thermo Fisher Scientific) invented a method called TaqMan (pronounced "TacMan" as a playful reference to the PacMan videogame), where a fluorescent signal would be given off by the activity of the polymerase as it simultaneously synthesized the complementary strand while degrading a specially labeled DNA probe.

Through careful monitoring of the amplification while the reaction was taking place, scientists could extrapolate an approximate concentration of the original target molecule. This would be much more useful than a positive or negative result of the PCR process as it would give researchers the relative number of molecules present from the sample.

This fluorescent signal could be detected through an optically clear plastic, which sat inside a metal block heating and cooling for each amplification round of DNA strand separation, the primer binding to a specific DNA location on each of the opposite strands, and the polymerase continuing its job of elongating the respective complement strands. The DNA probe would be simultaneously degraded, releasing a fluorescent molecule picked up by a laser. The optical signal would be analyzed, and as a function of the amplification cycle and comparison to standards, the relative concentration of original molecules could be determined.

This relative measurement of molecules is important to determine the viral load — the number of infectious disease particles of the individual being tested. In the case of the novel coronavirus, there has not been a clear relationship between the amount of virus an individual may have and the severity of their COVID-19 disease. How infectious particular individuals are as a function of the amount of virus they have is still an active topic of investigation.

The advent of what is called "real-time PCR" enabled key insights into the HIV epidemic; not only could the evolution of HIV be tracked and traced in real-time from individual to individual, but the amount of HIV in a person's bloodstream could be assessed with precision and their therapy adjusted accordingly. All of this came with FDA-regulated equipment and reagents, to ensure products intended "for research use only" would not be used to diagnose or treat human disease.

In the case of HIV (and also SARS-CoV-2), performing cell cultures and getting a live viral titer is time-consuming work that requires a specialist's expertise and the availability of special cell cultures, not to mention special biosafety precautions put in place to prevent worker infection. Real-time PCR technology, a standard workhorse in the molecular pathology laboratory, is performed routinely in many hundreds (if not over a thousand) of testing laboratories worldwide.

The real-time PCR business grew to over a billion-dollars, and this was before the COVID-19 pandemic. In a later section, we will take a closer look at the commercial incentives to examine alternatives to real-time PCR, given the practical limitations of the technology.

Next, however, we will look at one of the crowning achievements in biology: the Human Genome Project, which started in 1989 as a 15-year, $3.8 billion project sponsored by the U.S. Government. It finished ahead of schedule in 2001, adding 310,000 jobs and $796 billion in economic impact and giving impetus to a number of technologies we will cover here being put to active use in tackling the COVID epidemic.[10]

TECHNICAL ADVANCES IN SEQUENCING

In the 1980s, the length of a contiguous sequence (a "sequence read") that could be read on a given instrument was up to three or four hundred bases long.

The lengthening of DNA sequence reads was a result of progressively better enzymes, more creative biochemical mixtures to set up optimal reaction conditions, and better methods of separating out the G-A-T-C signals using a progression of easier-to-use methods. During the Human Genome Project throughout the 1990s, the length of an individual sequence read was more than double what had been possible in the 1980s, equivalent to six to eight hundred bases.

In the 1980s, for those of us who did this as a routine part of our duties, the process required a fair amount of careful work. Setting up the sequencing reactions was a matter of carefully pipetting microliter volumes and then mixing and incubating them in a PCR thermal cycler machine. The polyacrylamide gel was poured (before it hardened) between two glass plates that were 30cm by 40cm in size (about 12 inches by 16 inches) and held only 0.5mm apart; it was a challenge to clean these large glass plates in such a way as to prevent air bubbles from forming, as that would ruin an entire sequencing run. After several hours of electrophoresis (running electrical current across the glass plates in the presence of a buffer solution), the gel had to be separated from the thin plates, applied to paper, dried down under a vacuum using a special apparatus, and exposed to an X-ray film in the presence of a special cassette in order to amplify the radioactive signal.

The tedious part was reading out the sequence signal from the X-ray film (called an autoradiogram) and manually typing in what G-A-T-C base was indicated on the film to get it into the computer. The read-lengths of sequences at the longer end grew progressively harder as the bands became closer and closer together. It was a matter of faith in your own ability to ensure that the sequence was entered accurately; it was difficult to convince a coworker to double-check your eight-set of sequence reads at 400 bases apiece.

In the 1980s, fluorescent technology — out of the California Institute of Technology (CalTech) — applied to DNA sequencing led to the founding of Applied Biosystems in Foster City, California, and in 1986, they announced the world's first commercial sequencer, which was called the Applied Biosystems Prism 370A DNA Sequencer.

If you were diligent, six samples of purified DNA in the morning meant six sequencing reactions; about eight hours later, you would put the X-ray film in the cassette for an overnight exposure in the -80C freezer (the special plates used to amplify the radioactive signal worked best under low temperatures), and the next morning, you would have six lanes of 400 bases of sequence, or 2,400 bases of data, which had to be tediously entered into the computer by hand.

Remember, the human genome is 2.89 million pages. About thirty years ago, it took a full day and a half of effort to get to a single page.

The Applied Biosystems automated sequencer took a major step forward by automating the readout with fluorescent dye instead of radioactivity. While the gels still needed to be poured and the PCR-mediated sequencing reaction still had to be mixed and processed, the input into the computer was no longer the tedium it had been before. This was a leap in overall productivity that led to a corresponding lowering of the cost per DNA base sequenced.

Less than 15 years after the launch of the Abi Prism 370A DNA Analyzer, several advances in format and throughput enabled the Human Genome Project: the tedious gel-pouring was replaced by hair-thin capillaries filled with a separation polymer, the proprietary functional equivalent of acrylamide; the read lengths became longer with improved dye chemistry and now reached 600 or even 800 bases; lastly the number of capillaries and speed of sequence were increased — the Abi 3700 had 96 capillaries at once and could finish a sequence run in three hours.

Sequencing DNA was rapidly becoming better, faster, and less expensive.

CATALOGING VARIATION AFTER THE HUMAN GENOME PROJECT

Two decades after the completion of the Human Genome Project, the work of translating the basic discovery of human genes and their function into new medicines and diagnostics continues at a rapid pace. One last area to cover in the context of the Human Genome Project is the development and refining of another genomics tool: the microarray.

Now that the human genome reference had been completed, individuals could be compared to that reference, and it was microarrays that offered an efficient way to measure this variation.

Microarrays are an inexpensive and time-saving tool designed to show how individuals differ from each other in genetic terms. Instead of the entire genome (those 2.8 million pages and 520 miles of letters), you have specific single bases spaced throughout the genome. Depending on the question to be answered, microarrays are faster and much cheaper to perform compared to sequencing the entire genome.

The consumer genetics firm 23andMe and the genealogy firm Ancestry.com use these genetic microarrays, and in a later section we will discuss what they and other medical researchers have discovered in relation to COVID-19 genetic susceptibility.

THE ADVENT AND USEFULNESS OF MICROARRAYS

Genotyping is the method of determining genetic variation over specific nucleotide bases that naturally occur in the human genome. At their most global impact, these variations in DNA are what make each of us unique. At their most acute, it is Single Nucleotide Polymorphisms (known as SNPs and pronounced "snips") that vary by population and individual. These SNPs, whether in the 1% of the coding genes in the genome or the 99% of the remainder, are markers that can be seen as similar to signposts along the Appalachian Trail.

The impetus for researching human variation is clear: to determine the underlying genetic contributor to disease.

While sequencing equipment was becoming automated, the microarray was invented and developed at Stanford University and then commercialized by the companies Agilent Technologies, Affymetrix (now part of Thermo Fisher Scientific), and Illumina Incorporated.

By labeling a biological sample and then applying that DNA to a special glass slide or genetic chip, hundreds of thousands of genetic variants at specific locations throughout the genome could be measured with great accuracy. For as little as $199, you can determine a fair amount of information from these measurements. As is well-known, you can find out what percentage of a particular ethnicity you have. You can even find out the percentage of Neanderthal DNA. Entertaining information such as whether your earwax is wet or dry or whether you have a natural sensitivity to bitterness in kale-like foods is available with a few keystrokes.

The amount of medical information to be obtained from these genetic microarrays is limited and has to be handled by qualified medical personnel such as a genetic counselor or a physician. Some *BRCA* breast cancer susceptibility information is there, albeit limited in scope when compared to the sequencing of the entire gene family. Other susceptibilities, such as the likelihood of developing Alzheimer's Disease through the genotypes of the *APO-E* gene, are understandably highly consequential, and this information needs to be handled with care.

These microarrays were developed in the early 2000s to analyze genetic variation; the synthetic DNA on the chip itself was functional DNA and could be used as a sequencing primer — as a starting point for the DNA polymerase to do its work of synthesizing a complementary strand. In the case of microarrays, you are sequencing a single base at a specific location.

Fast-forward to less than 20 years later, and 23andMe has analyzed its own database of over a million research participants and asked them to volunteer their own COVID-19 diagnosis along with the severity of their disease. They discovered that Type O blood is a protective characteristic against testing positive for SARS-CoV-2, lowering the chance of infection by some 9-18%[11]. These preliminary findings will be supplemented with additional genetic insights as more data is collected.

Comparing the early 2000s to today, the ease of collecting and analyzing vast datasets is unparalleled. Next, we will cover a key breakthrough in genomic technology, which came after the Human Genome Project was completed and has greatly accelerated the pace of genomic discovery yet again. This capability enabled researchers in Shanghai to sequence the genome of the novel coronavirus in a very short amount of time, publishing the 29,903 bases on January 10, 2020.

This key breakthrough in genomics is a new method of sequencing called Next Generation Sequencing (NGS).

Its current importance in the context of COVID-19 has been to allow scientists to track individual strains and their propagation throughout the world over time and trace whether or not there are differences in infectivity or COVID-19 severity in a strain-dependent fashion.

An emerging possibility for putting this technology to work has been its potential to increase the volume of diagnostic testing many times over, and it may play an important role if screening should become commonplace in 2021.

CHAPTER 4: NGS A MIRACLE OF TECHNOLOGY

Everything is hard before it is easy.

— Brian Tracy

In 2003, the state-of-the-art sequencing instrument built by Applied Biosystems was called the 3730xl and had 384 hair-thin capillaries. If it ran 24 hours per day as it was designed to do, it could sequence one million bases per day.

This was the same Applied Biosystems that built the first automated system in 1986, then the workhorse instrument for the Human Genome Project in 1998.

In 2005, a new company called 454 had a new approach to sequencing that was radically different: instead of 800 bases of DNA being sequenced in a stretch, it would only sequence around 100 bases. What it did not have in information length, it more than made up for in information width. The 454 GS-20 FLX produced about 100 times more data per 24-hour day than the ABi 3730xl — or, 100 million bases per day.

In 2006, Illumina CEO Jay Flatley asked for an all-employee teleconference. He announced the acquisition of a start-up out of Oxford, U.K. called Solexa, which had developed a system that would leapfrog the 454 system: it would increase the daily throughput to 300 million bases, an increase of threefold.

AN EXPLOSION OF CAPACITY AND A COLLAPSE IN COST

Thanks to healthy market competition and demand across research, pharmaceutical, and clinical markets, it is estimated that there was a total market size of $5 billion in 2020, up from zero in 2005. Since then, the early leapfrog in capacity and lowering costs has only continued at a blistering pace.

In 2008, the first 454 complete individual genome was sequenced; it was the DNA of the co-discoverer of the structure of DNA, James Watson, completed at a cost of $1,500,000. The second individual was one-thirteenth the cost of the first individual — about $110,000. In 2012, using the Solexa technology Illumina called the HiSeq 2000, the cost had collapsed to $10,000, one-one-hundred-and-fiftieth the cost of the 454 technology only four years before.

By 2016, that cost had decreased another 10-fold to the $1,000 genome. Remember the 520 miles of printout? All that data could now be produced at 1/1,500th the cost of what it had cost eight years before.

This is hard to comprehend, as we do not often come across such a dramatic decrease in the price of something. Imagine a $65,000 car being offered to you for $43 — this is the magnitude of improvement we are talking about.

Much has been made of Moore's Law, which the founder of Intel stated as the density of transistors doubling every eighteen months to two years, with a concomitant decrease in computing power. This law, stated in 1985, has proven to be remarkably prescient.

Figure 6 shows Moore's Law compared to sequencing costs tracked over a 19-year timespan.[12] Over this longer time-frame, the decrease is some 33,000-fold!

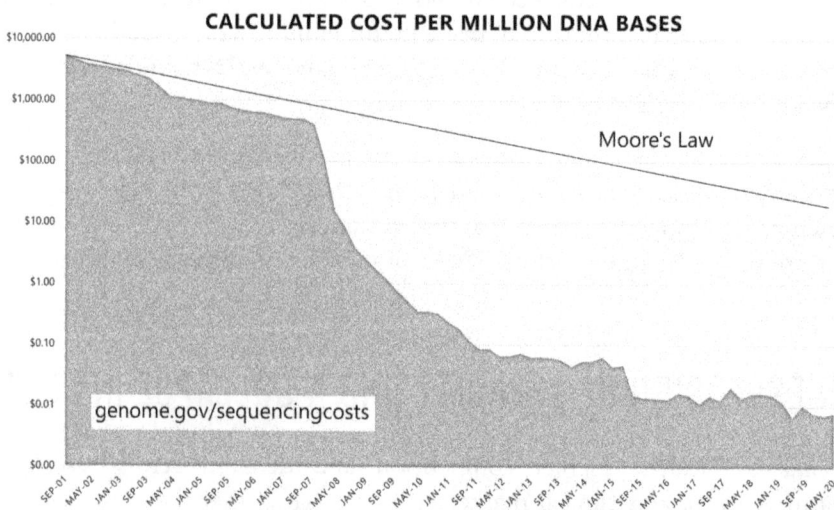

CALCULATED COST PER MILLION DNA BASES

Figure 6: Decline in the cost of sequencing over two decades. Moore's Law compared to collapsing prices per million bases of sequence information, driven by Next-Generation Sequencing (NGS). The scale on the Y-axis is logarithmic, showing a 33,000-fold decrease over the course of 19 years.

APPLYING NGS TO THE NOVEL CORONAVIRUS

Though NGS is used extensively by researchers worldwide, many questions remain unanswered. One is genetic susceptibility (which we will cover in a later section), which may lead to clues on new therapeutic approaches. Another is real-time monitoring of virus mutations and whether these individual mutations of the virus make a particular strain more infectious or cause an increase in the severity of COVID-19 disease.

A third question is why this virus is so much more virulent and transmissible than the SARS or MERS outbreaks of 2002 and 2009 respectively. To that end, researchers are actively looking at many dimensions of biology, from the expression of *ACE2* receptors on different organs to the nuances of the immune response (or lack thereof) among individuals. These research frontiers heavily depend on NGS technology in order to carefully understand the biology of the disease.

On December 12, 2019, the first patient was hospitalized with the novel coronavirus in Wuhan, China; by January 10, 2020, a full 41 people had been infected. On that same day, a group of scientists in Shanghai uploaded the 29,903 bases of what they then called novel coronavirus-2019 to an open-access database used by virologists for pre-publication data, Virological.org.[13]

They did this in record time thanks to NGS instruments confirmed using Sanger sequencing.[14]

Many other advances in technology enabled global data sharing: computers to analyze this data, the global Internet, a public database to deposit this data in, and electronic communications of all types.

THE THREE WEAPONS IN THIS WAR

Having the entire genome sequence of this virus available so quickly enabled the three weapons in this war against the coronavirus to be activated all at once. With the sequence in hand, synthetic DNA primers and probes for PCR diagnostic tests could be designed and ordered for manufacture; the coding region for the spike protein (a key protein in the entry of the virus into a vulnerable cell) could be synthesized to express the protein for biochemical and structural studies, which was essential for drug development; and new companies with DNA or RNA-based vaccines could embark on a record-setting sprint to produce a vaccine against a disease that humanity had not seen before.

Of primary importance here is diagnostics. Primers for real-time PCR could quickly be designed, synthesized, and tested in the laboratory for performance and accuracy. Anyone with a computer connected to the Internet could download the sequence of this novel coronavirus, design a set of PCR primers, and double-check it for specific amplification of that virus rather than any other respiratory virus or coronavirus, as well as order primers and fluorescent probes and start the hard work of developing a clinical test.

On the therapeutic front, the coronavirus genome's components could be analyzed and experiments designed to better understand how the virus entered the body. The first and primary analysis showed nCoV-2019 (later called SARS-CoV-2) had the closest sequence similarity to SARS-CoV and a distant relationship to MERS-CoV. Specifically, the virus has a seldom-mutating protein called the spike protein that interacts with *ACE2* receptors on human cells that line the upper respiratory tract.

Biologists would take the sequence of the spike protein; engineer it into cells to produce this viral protein (the cells could be bacterial, insect, yeast, or mammalian in culture depending on a number of considerations); and purify large quantities of this viral protein in order to obtain the X-ray crystal structure of this key protein.

The work would also involve co-crystallization with human *ACE2* receptor protein and even co-crystallization with and without drugs in development to interrupt the interaction, thus barring entry of SARS-CoV-2 into the cells lining the human airway.

On the vaccine front, companies that already had development programs underway — using new methods of vaccination for influenza or other coronaviruses such as MERS — immediately used the SARS-CoV-2 sequence to begin designing a COVID-19 synthetic RNA-based vaccine. Other companies used the viral sequence to begin the other four approaches to vaccine development. These included engineering the specific sequence of the virus into a suitable adenovirus designed to infect human cells and have those cells produce, but not replicate, the coronavirus protein.

Other companies and academics (notably a group in China) would work to produce vaccines using more established methods which, instead of being nucleic acid-based, were based on immunizing against a specific viral protein. By using the sequence of the viral protein, scientists would insert this sequence into an insect or mammalian cell culture in the laboratory and produce a purified spike protein to use as a vaccine.

Of the almost 200 vaccines in development, there are five basic methods being used, and none of them would have been possible without the sequence of the viral genome.

NGS has the potential to vastly expand the capacity for viral diagnostic testing far beyond what real-time PCR can achieve. Already, NGS has transformed cancer diagnostic testing for solid tumors and blood cancers and is going through a further transformation by enabling cancer diagnostics through a simple blood draw, which is called colloquially a "liquid biopsy" (compared to an invasive biopsy).

We will come back to NGS in later chapters, but we now turn to PCR-based diagnostics, the first front in this war — the fire alarm without which we are completely blind.

CHAPTER 5: BATTLING AN EPIDEMIC THROUGH MOLECULAR DIAGNOSTIC TESTING

I approached him in a humble spirit: "Mr. Edison, please tell me what laboratory rules you want me to observe." And right then and there I got my first surprise. He spat in the middle of the floor and yelled out,

"Hell! there ain't no rules around here! We are tryin' to accomplish somep'n!"

And he walked off, leaving me flabbergasted.

— **Martin André Rosanoff (From "Edison in his Laboratory" 1932)**[15]

I arrived at Dulles International Airport's Arrivals area right on time. It was January 31, 2020, and my daughter's excitement over studying abroad in Beijing had just been cut short by a hasty return. While filled with disappointment that a virus from Wuhan would so disrupt her plans, I had optimism that progress would be made quickly. Checking my phone's newsfeed while waiting, I read the CDC's latest update: The first real-time PCR kits to detect "novel coronavirus-2019" were close to FDA authorization and were scheduled to go out the following week.

February of 2020 was a highly consequential month for the U.S. Conflicting messages from many sources, compounded from both mass media and social media, led to confusion building alongside justified concern about the drumbeat of ominous news from China. The 20 million residents of Wuhan were completely locked down. Arresting images were shared of streets blocked by dump trucks depositing mounds of dirt, which were meant to block any exit or entrance into the city. Authorities went so far as to weld people into their apartments. News of outbreaks in other cities in China and across Asia was not reassuring.

In hindsight, accurate information would have made a world of difference. With this highly transmissible disease, which has serious public health consequences, a few key points now stand out: Masks reduce the spread, physical distancing works, and infection by pre-symptomatic and asymptomatic people makes controlling the virus particularly hard. Furthermore, regardless of the quality and accuracy of the diagnostic test and its underlying technology, there will inevitably be false negatives thanks to the nature of the disease.

The nature of the symptoms needs to be examined before an examination of diagnostic testing, as symptoms can confirm a diagnosis even when an individual has tested PCR-negative.

CLUES FOR DIAGNOSING COVID-19

At the beginning of this new disease, the collection of symptoms varied greatly and were dependent on the individual. The list of COVID-19 symptoms from a review of hospitalized patients was as follows: The most common symptom was a fever (70%-90%); followed by dry cough (60%-86%); shortness of breath (53%-80%); fatigue (38%); muscle pain called myalgias (15%-44%); nausea, vomiting, or diarrhea (15%–39%); and headache or weakness (25%). Laboratory blood tests include readings of lymphopenia or low immune cell blood count (83%) and elevated inflammatory markers of several types (46%).[16] CT scans often reveal a "ground-glass opacity" representing fluid in the lungs, which was at first thought to be a hallmark of infection, but later ruled out as too inconsistent.

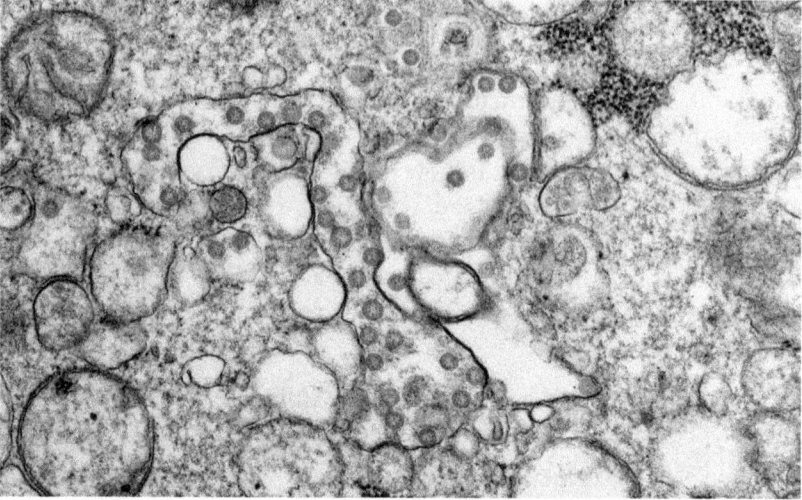

Figure 7: Transmission electron microscopic image of an isolate from the first U.S. case of COVID-19. The spherical viral particles (appearing as small dark circles) contain cross-sections through the viral genome, seen as black dots. Public domain image from the U.S. Centers for Disease Control, Hannah Bullock and Azaibi Tamin.[17]

These CT scans have very high sensitivity (low false-negative rates) in catching more cases of SARS-CoV-2 infection and COVID-19 disease. However, they have very poor specificity (high false-positive rates) due to the imaging results being similar to those resulting from other viral infections. To confirm a positive diagnosis, a combination of symptoms, imaging, and molecular testing is used.

On top of these symptoms are common and serious complications for those patients who must be hospitalized. Pneumonia (75% of those hospitalized) is the most common complication, followed by acute respiratory disease syndrome (15%), acute liver injury (19%), cardiac injury (7%-17%), and damage to several other organs, including kidney injury and neurological damage.

DIAGNOSTIC TESTS LAG BEHIND THE VIRUS

Molecular diagnostic testing here signifies the detection of the coronavirus RNA. In the research lab, the "gold standard" of actually growing virus from a patient sample in a susceptible cell culture is possible. However, this task is cumbersome, time-consuming, and expensive to perform, requiring specialized knowledge and equipment. It is not practical to perform it on a wide basis. A PCR test is the feasible route.

Bear in mind that PCR testing measures the presence (or absence) of coronavirus RNA molecules, rather than detecting infectious virus particles. The current recommendation is to take the relative number from the PCR test in combination with clinical symptoms.[18]

There is evidence that individuals can continue to test positive for the presence of coronavirus molecules long after they cease being infectious. Figure 8 illustrates the probability of detection by week, moving from time of exposure (area on the left showing the time of two weeks to one week before symptoms appear). You can see that the solid line where you are able to transmit the infection to others is in the week prior to symptoms appearing and in the week after. Indeed, it is estimated that a full 62% of infected individuals have been infected by individuals who were in this pre-symptomatic phase.[19]

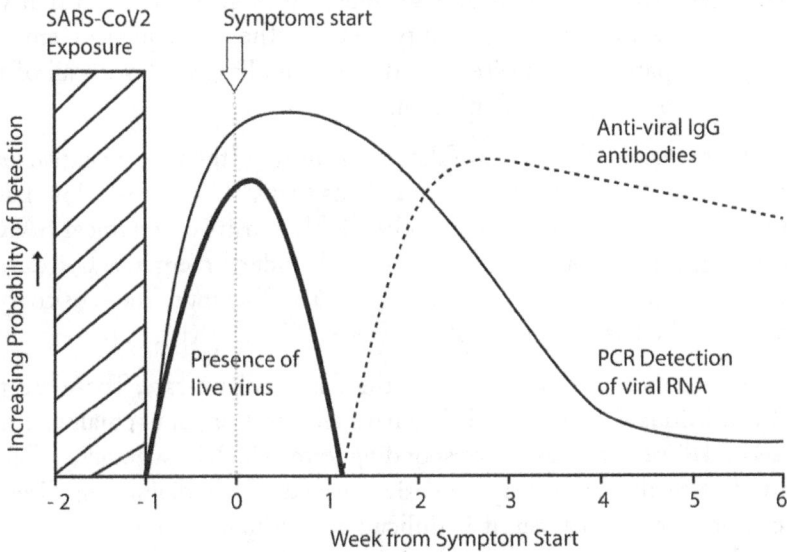

Figure 8: Probability of detection by week from symptom start. The bold line is live virus isolated from the respiratory tract, the thin line is PCR detection, and the dashed line is IgG antibody detection. This is an estimate and the values should be considered approximations.

The spread of disease while still pre-symptomatic is the real problem here, which is why the non-pharmaceutical interventions of wearing masks, maintaining physical distance, and awareness of the availability of fresh air are of paramount importance. In a public service announcement in Japan, they have a campaign called "Avoid the Three C's: Closed spaces with closed ventilation, Crowded places with many people nearby, and Close-contact settings such as close conversations."

Public confidence in the usefulness of tests for COVID-19 has also been called into question. The challenge is not the quality of the tests, but how these tests are interpreted.

WHAT IS A POSITIVE DIAGNOSTIC TEST ANYWAY?

Let us take a look at a basic understanding of what truth is in terms of what is deemed "positive" or "negative" as a test result, using the diagnosis of syphilis as an example.

A diagnostic test has far-reaching consequences. An incorrect result which is a false-negative (absence of disease when infected) will tell you that you do not have syphilis when in reality you do. The consequences are far-reaching: The patient has a disease and it goes undiagnosed, with all of the consequences of undiagnosed infection.

The first dimension of accuracy of diagnostic testing (the false-negative rate) is described using the diagnostic term "sensitivity." If there is a low false-negative value, the test is deemed to be highly sensitive. A typical SARS-CoV-2 diagnostic test authorized by the FDA under Emergency Use during this epidemic has a false-negative rate in the 1%-2% range. The sensitivity is 100% minus the false-negative rate, or 98%-99% sensitivity.

The second dimension of accuracy is the false-positive rate. These are the healthy individuals who are told they have an infection, but actually do not. This second dimension has a corresponding term, which is "specificity." False-positive test results can be very expensive mistakes when diagnosing diseases like cancer. For lung cancer, it is difficult to distinguish between a benign lung nodule and a cancerous one by imaging; absent of other symptoms, a lung cancer diagnosis requires surgery in order to get a biopsy performed, which entails its own risks.

Similar to sensitivity, specificity is 100% minus the false-positive rate. For SARS-CoV-2 diagnostic testing, the specificity range is also 98%-99% specificity.

With the highly transmissible and potentially fatal COVID-19 disease, the sensitivity of the test (a low false-negative rate) is much more important than its specificity (potential for high false-positives) from the perspective of curbing a runaway epidemic.

How tests are developed and regulated to ensure a highly accurate test, with low false-negative and low-false positive performance, will be examined next.

REGULATION OF CLINICAL LABORATORIES

Laboratory managers and directors not only worry about the important test performance details of false negative and false positive rates, but also about the potential for cross-contamination, sample mix-up, and ensuring consistent test performance over time (the federal agencies, when evaluating laboratory tests, do so as well). There exists a welter of federal regulations to ensure laboratory procedural quality, known as the Clinical Laboratory Improvement Amendments of 1988, Public Law 100-578, which were further modified in 1997 and 2012. Another set of major amendments are currently working their way through Congressional committees.

The regulations ensure that these laboratories have appropriate training, written protocols, and procedures, as well as systems for maintaining ongoing quality checks.

CLIA involves three Federal agencies: the Food and Drug Administration (FDA) (which categorizes tests based on their complexity and adjudicates "CLIA waivers," exemptions from CLIA regulations); the Centers for Medicare and Medicaid Services (CMS) (which issues certificates, conducts inspections, manages private accreditation, and approves proficiency testing [PT] conducted at the state level); and the Centers for Disease Control and Prevention (CDC) (which develops technical standards, develops laboratory guidelines, monitors PT practices, and conducts quality improvement studies).

CLIA is necessarily complicated. For the clinical laboratory, CLIA certification signifies whether or not you are able to run a business that handles patient samples and issues results.

In the Theranos debacle, a high-visibility and highly-valued start-up ran faulty clinical tests under a loophole in the CLIA guidance called the Laboratory Developed Test (LDT). The LDT is a diagnostic test that meets the FDA definition as "designed, manufactured and used in a single laboratory." The reason for the existence of the LDT is to enable innovation that will drive the growth of new tests in the marketplace. One market research firm estimates the market for LDTs in 2020 to be $12.8 billion dollars.[20]

A FAILURE OF THE REGULATORY SYSTEM

In the last chapter, it was noted how quickly the genome of the SARS-CoV-2 virus was sequenced; it was made available to the world's scientists on January 10, 2020. Only ten days later on January 21, a group in Berlin submitted a research paper describing their testing method and protocol.[21] The CDC announced on the same day that they had finalized their test and were working on distributing it to public health laboratories.

On February 4, 2020, U.S. Secretary of Health and Human Services Alex Azar announced a public health emergency. Recognizing the potential for abuse of the LDT system, in that same announcement, the FDA specifically prohibited clinical laboratories from launching their own LDT for the detection of SARS-CoV-2, announcing the Emergency Use Authorization mechanism for diagnostic testing for COVID-19. At the same time, the FDA announced that the CDC's test was approved under EUA guidance underneath this public health emergency.

The process of applying for an EUA required documentation that any laboratory running an LDT would be familiar with. This regulatory hurdle for an EUA, however, was a newly invented process that not only eliminated the potential for poor tests to be launched into the marketplace, but also prevented very well-qualified and expert laboratories from launching their own tests to get ahead of early infections.

The insertion of a regulatory hurdle for the existing laboratories to develop their own LDTs, as well as the only EUA approval for the CDC test having a reagent issue, caused precious weeks to be lost in February of 2020. Existing clinical laboratories with the expertise to run tests were additionally hamstrung in the CDC's requirement to have multiple viruses tested for simultaneously, as this required precious patient samples from prior coronavirus outbreaks from 2002 and 2012. For reasons that are yet to be explained, the CDC decided their coronavirus test would also test for MERS (the coronavirus from 2012 to present) as well as for SARS (from 2002 to 2004). These laboratories were hobbled by a need to have positive samples that no laboratory could locate.

On top of this onerous requirement, the tests the CDC sent out were faulty, as shown by laboratories trying to implement these tests. One of the chemical reagents had a contamination issue, and it was not until February 28, 2020, that the test was revised (only testing for the novel coronavirus, not for MERS or SARS); the EUA process was clarified and test developers could proceed. Two weeks later in mid-March, the first EUA outside of the CDC was approved, with another seven approved in early April. It is widely accepted that the loss of early testing (the CDC estimates only 200 samples were run in the entire U.S. in March, in stark comparison to other countries) contributed greatly to the spread of the epidemic in the U.S.[22]

It is also important to note that even if the FDA authorizes a coronavirus test under EUA, it does not simply mean that a laboratory can obtain a test kit and start collecting, running samples, and returning test results to patients. The clinical laboratory has to first perform something called a test validation.

VALIDATION OF A CLINICAL TEST

Validation ensures that a test measures what it is supposed to measure, specifying its sensitivity and specificity for detecting SARS-CoV-2, as well as other measures such as the "Limit of Detection" (LoD — how few virus RNA molecules it can pick up and call a positive result) and to what level the test is affected when other viruses or bacteria are present.

For example, what if a person has both the flu and the novel coronavirus? Will that affect the result? If the patient's nasal swab sample also has some cough medicine in it, will the result be affected? If pure water is used as the sample, will the test pick up contaminating molecules from the laboratory and show a positive result? These are important considerations.

Validation is a lot of work; several dozen positive patient samples and negative controls have to be sourced, procedures are written, training established, quality methods established, and a Quality Management System also has to be established to document important qualities of a test. Every CLIA laboratory has a director responsible for all this activity who are among the true heroes during this pandemic.

The validation process involves a set of carefully controlled experiments that determine true-positive and true-negative rates (sensitivity and specificity respectively), as well as the aforementioned limit of detection. Confusingly, this is also often named sensitivity, as it determines how few molecules of virus RNA (in absolute terms) the test can reliably detect. As a reminder, these molecular tests measure an approximate number of coronavirus RNA molecules, not the live, infectious virus itself.

Limit of detection is a tricky topic, especially for PCR-based tests that are able to detect a small handful of molecules. As a thought experiment, say you have a large bowl of jellybeans; half red and half white. If you grab a handful of 20, how many will be red? Ten is the most likely number, followed by 9 or 11, then followed by 8 or 12. After multiple grabs at the jar, you'll be able to calculate the average percent of red jellybeans that were in the jar and what the range of red jellybeans was from a set number of handfuls.

In statistics, this is called a "confidence interval." In the red jellybean example, you can have an average of 50% red jellybeans and a confidence interval of 95% between 49% and 51%. For any handful of 20 jellybeans, 95% of the time the percent of red jellybeans would fall between 49% and 51%. For LoD measurements and for sensitivity and specificity measurements, a number and a confidence interval would also be calculated.

In normal times, validation studies are published in the scientific literature before the FDA sees them in an application package, but during a public health emergency, the validation documentation is made a part of the Emergency Use Authorization (EUA) submission in order to bring a new test to the clinic and the market more quickly.

THE FDA EMERGENCY USE AUTHORIZATION

The language around EUAs is very clear: These tests are authorized specifically for the duration of the SARS-CoV-2 pandemic (under Section 564 of the original act of Congress called the Food, Drug, and Cosmetic Act)[23]. Do note that these are not FDA Approvals, these authorizations are a temporary measure. It is also noteworthy that a key amendment to Section 564 is called the Pandemic and All-Hazards Preparation Act of 2013.[24]

Remember, the CLIA regulations do not set standards for diagnostic tests, they only ensure that the documentation is in place, the validation requirements are met ("it does what they say it does"), and it is being run by a licensed CLIA laboratory. Notably, such a test is also much more likely to be reimbursed by Medicare or a private insurer.

FDA STANDARDS VS. CMS STANDARDS

In the United States, the FDA will approve a lab test if it is "Safe and Effective" (the same standard as for medicines). Under the ordinary FDA approval process, clinical diagnostic tests go through a clinical trial process as well, although a Phase I pilot trial for safety is not needed. More invasive tests (such as a laparoscopy) require much more clinical trial data on the "safety" side in order to win FDA approval.

In the United State's public/private hybrid system, CMS has assumed a key role in determining what tests get paid for and to what extent. A hodgepodge of eleven private subcontractors is involved in determining what is "Reasonable and Necessary." This is a higher standard than "Safe and Effective" — just because a CLIA-certified lab is running a diagnostic test does not mean that it will be paid for by Medicare. The test must be vetted and approved by a CMS subcontractor first.

Reimbursement is a price-setting mechanism that plays a major role in determining the profit a clinical laboratory might make on any given test. CMS reimbursement rates and corresponding billing codes (known as CPT® "Current Procedural Terminology" codes; five-digit numbers published by the American Medical Association) are what need to be established in order for a test to be seen by the medical establishment as "Reasonable and Necessary."

In medical practice in the United States there is a concept called "Standard of Care." This signifies what a doctor should do when given a specific set of symptoms and information from diagnostic tests. For COVID-19 disease, the Infectious Diseases Society of America has an often-updated set of guidelines ("Guidelines on the Diagnosis of COVID-19")[25] with 15 diagnostic recommendations, specifically around "Nucleic Acid Amplification Tests."

Their first recommendation reads as follows:

> *The IDSA panel recommends a SARS-CoV-2 nucleic acid amplification test (NAAT) in symptomatic individuals in the community suspected of having COVID-19, even when the clinical suspicion for COVID-19 is low (strong recommendation, very low certainty of evidence).*

DALE YUZUKI, M.A., M.ED.

Remarks:

- *The panel considered symptomatic patients to have at least one of the most common symptoms compatible with COVID-19 (Table 1).*

- *Clinical assessment alone is not accurate in predicting COVID-19 diagnosis.*

- *The panel considered timeliness of SARS-CoV-2 NAAT results essential to impact individual care, healthcare institution, and public health decisions. In the outpatient setting, results within 48 hours of collection is preferable.*

You can see that results within 48 hours are "preferable," although most tests take longer, which led Bill Gates to say in an interview that most tests in the U.S. are "complete garbage" due to the speed of transmission and asymptomatic spread.[26]

The strains on the existing testing system are manifold as the U.S. tries to scale up the volume of tests to tens of millions per month; the methodology for most PCR tests is the following:

- Nasopharyngeal swabs are placed into 3 mL (about a teaspoon) of a liquid called Viral Transport Medium.

- RNA purified from the sample in the medium.

- Set-up and incubation of a preliminary reverse transcription reaction to convert viral RNA if present into DNA.

- A quantitative real-time PCR test run, including positive and negative controls.

- Software analysis reporting a positive or negative, along with a measure called Ct (Ct is the number of PCR cycles where the amplification signal passes a preset threshold).

This presupposes the availability of swabs, transport medium, RNA purification kits, and qRT-PCR (quantitative real-time PCR) reagents for the detection of the SARS-CoV-2 viral RNA. Also assumed in the example above are systems designed to label each sample with a unique identifier (typically a long string of numbers and letters and its related barcode) and an information system to report results back to the physician and patient.

Add to this that the availability of the Personal Protective Equipment (PPE) required for the Clinical Laboratory Technician and all other components (not to mention qualified and trained personnel) has been an enormous challenge. It remains a major challenge to increase volumes another 10-fold or 100-fold.

Considering the nature of the pre-symptomatic spread of the disease, the major health impact of false-negative results, and all the work that goes into the development and validation of a clinical test, what are the real-world detection rates of SARS-CoV-2? And is there anything that can be done to improve false-negative results from PCR tests?

FALSE NEGATIVE RESULTS FROM PCR TESTS

One study published in March of 2020 looked at over 200 patients, where many different sample types were taken over the course of their disease — sputum, nasopharyngeal swabs, bronchial lavage (this is where they literally put liquid down your bronchial airway, which is certainly not a comfortable experience), stool, blood, and urine.[27]

What this report discovered was astonishing: a 10,000-fold difference in the amount of virus from the patient's nasal swabs, even though all of them clearly had COVID-19 disease. These patients were in the hospital, sampled within one to three days after admission, and only 63% of them tested positive by nasal swab; there was a full 37% false-negative rate.

Think about that for a moment: Of those admitted to the hospital to treat their COVID-19 disease, over a third showed up as negative for the SARS-CoV-2 virus in their upper respiratory tract.

Sputum (what patients will cough up) was a little better at 72% (a 28% false-negative rate), and lung lavage (wash fluid) was best at 93% (a 7% false-negative rate).

A PCR-based molecular test is typically very sensitive. A laboratory has to determine their test's limit of detection, often via the 5-15 viral RNA molecules per milliliter of sample fluid. For specificity, the laboratory determines the ability of their test to pick up these small numbers of target viral molecules against a background of millions of copies of an assortment of other viruses.

The patients in this study had a battery of samples taken at the same time and molecular PCR tests were applied, so it was a fair comparison across sample types, yet testing still missed some patients with COVID-19 disease.

From a meta-analysis (a study of several studies) we can analyze the false-negative rate across seven independent nasal swab sampling reports. The focus was on known SARS-CoV-2 infected individuals identified through contact tracing efforts, with RT-qPCR tests performed as early as one day after exposure.[28]

The chances of a false negative are summarized in the chart. As you can see, on the day the symptoms appear, there is a full 38% chance that an infected person will get a negative test result! Even after COVID-19 symptoms have been present for a few days, the odds of a test coming back negative are in the range of 20% to 25%.

Day of Infection	Chance of a False Negative SARS-CoV-2 Positive but PCR Negative	95% Confidence Interval Range
1	100%	100% to 100%
2	100%	80% to 100%
3	95%	50% to 100%
4	67%	27% to 94%
5 (Symptoms)	38%	18% to 65%
6	25%	15% to 42%
7	22%	11% to 33%
8	20%	12% to 30%
9	21%	13% to 31%
10	24%	15% to 35%
//	(gradual increase from 24% to 44%)	
15	44%	31% to 57%
//	(gradual increase from 44% to 66%)	
20	63%	52% to 73%

Figure 9: The probability of having a negative RT-qPCR test result even though you are infected with SARS-CoV-2 ranges from 20% to 38% from the day you have symptoms of being sick.

Think about that for a moment. The clinical laboratory — with all of its PCR technology and the ability to detect ten copies of virus RNA from a nasal swab — still reports a test that comes back negative?

How can the virus just disappear while the disease ravages a patient's lungs and other major organs? Is the mistake due to the mishandling of the sample from the time of collection or improper swab sampling by the healthcare worker? Or is something else at play?

This is the awful nastiness of this coronavirus. Without symptoms and a few days after exposure there is still excess virus to spread disease. Once symptoms appear, you have a now-you-see-it and now-you-don't molecular presence in the upper respiratory tract. A widely heterogeneous response among individuals, ranging from those who are completely symptom-free, to suffering from full-blown COVID-19 and possible death in a matter of days.

The evidence is now clear: Infected individuals are most contagious prior to symptom onset.

Getting a better false-negative rate for testing individuals for the novel coronavirus is possible with bronchial lavage. Among hospitalized patients with confirmed COVID-19, the false-negative rate is the lowest at 7% — or a 93% sensitivity of detection. However, obtaining bronchial lavage samples is highly impractical for several reasons, the least of which is the discomfort of the procedure. Impracticalities also arise from the need for specialized training, a hospital environment, and the equipment required to perform the procedure.

USING THE DIAGNOSTIC TOOLS WE HAVE

The novel coronavirus known as SARS-CoV-2 has a devastating combination of pre-symptomatic spread and a widely variable presence in the upper respiratory tract that makes molecular tests, for all their sensitivity and accuracy, much less effective in blunting the transmission of COVID-19.

The topic of transmission is next, where we will look at particular environments where closed, crowded, and close-contact spaces (the "Three C's") invite mass-spreading events.

CHAPTER 6: TRANSMISSION, SENSITIVITY, AND DIAGNOSTIC TESTING

In theory, theory and practice are the same. In practice, they are not.

— **Anonymous**

The CDC has estimated that the total number of Americans who have been infected is about ten times the number of known and confirmed cases. A fraction of this ten-fold number is made up of asymptomatic cases (those without symptoms). Looking at examples of well-studied outbreaks within indoor environments yields instructive lessons about how the virus spreads.

Of the early studies on transmission, one focused on a Korean call center stands out.[29] Researchers thoroughly studied this outbreak, where 1,143 people were tested, with 97 testing positive. The open-plan office on the 11th floor held 216 employees; 94 of the confirmed positive cases worked there.

The building is in the busiest area of downtown Seoul. Floors 1-11 house commercial offices, while floors 12-19 are residential. Prior to the outbreak, there was considerable interaction between workers on the 11th floor and the other floors of the building; in elevators, the lobby, and common areas. Researchers mapped where the infected individuals sat.

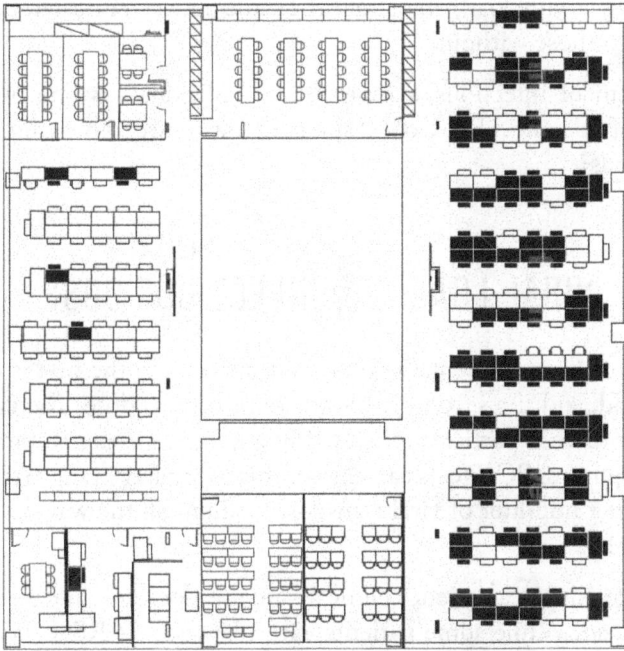

Figure 10: Floor map of the Korean call center. Infected individuals were in the shaded cubicles.

This data is remarkable: On one side of the 11th floor, 79 of 137 employees were infected. On the other side of the same floor, there were only a handful of cases. Infections spread over the course of two weeks before the building was closed. Employee's families were also tested; 34 of 225 households had family members who contracted COVID-19 as secondary infections from the call-center. The researchers followed up with all of the individuals for a full 14 days after the building was closed in order to detect all known cases from the outbreak.

This study yields some important practical lessons. By testing everyone who worked in the building (over 1,000), researchers determined that there wasn't a single known case resulting from casual contact, conversation in an elevator, or infection being picked up from a common area surface. The infection came from individuals sharing time in an enclosed space and speaking for extended periods of time. The number of asymptomatic cases in this outbreak was relatively low (only 4% of the total), and another 4% were pre-symptomatic (that is, symptom-free when tested, only to develop symptoms later). Thus a full 8% of individuals were infected and potentially infecting others without showing any symptoms. Later studies would indicate that the amount of virus (known as viral load) was about the same between infected people with symptoms to those without.[30]

This problem of infectious, symptom-free individuals is one of the most insidious aspects of SARS-CoV-2 spread. How do we know who is infected and infectious?

VIRAL LOAD AND INFECTIOUS DOSE

Two concepts from influenza virology are instructive: the first is one called the viral load, and the second is known as infectious dose. The viral load is the number of viral particles measured in a patient's upper respiratory tract or in their sputum. PCR tests can answer this over a large range (that is, from five copies per milliliter of viral transport medium all the way up to billions of copies/mL).

Perhaps surprisingly, a patient's clinical outcome does not directly depend on their viral load. While some patients with plenty of SARS-CoV-2 in their respiratory tract have had severe COVID-19, the course of disease for others has been mild.[31] To complicate matters further, some patients without any symptoms have had high viral loads.[32]

Scoring positive test results by the amount of virus is simply not very predictive.

The second concept to note here is the infectious dose as it relates to the body's exposure. You can have an individual with a high amount of virus, but you only spend a matter of seconds passing by that person at a walk. Another individual may only have one one-hundredth the amount of virus, but you spend ten minutes in their presence; your dose in this situation is much higher than in the former.

The minimum amount of virus needed to infect someone can be approximated by the viral load in the respiratory tract of the contagious individual multiplied by the time spent in their presence. You then need to take into account droplet and aerosol production, how close you were to that individual, and the effect of ventilation. To date, there is much to be learned about what the minimum dose for infection actually is; there is a ferret model for transmission, but it is not yet known how transferrable these findings are to humans.[33]

In the case of the Korean call center, the infection spread from one individual to dozens all along the same side of the office building. Close interaction and many hours of exposure to infected individuals was the common denominator among those infected.

A SECOND EXAMPLE OF TRANSMISSION

Another early incident that was important in understanding how SARS-CoV-2 spreads was an outbreak traced to a January 23rd dinner at a restaurant in Guangzhou, China. The "index case" was seated at a table in the middle of the dining area. Over the course of 90 minutes, four of the other nine people at this table were infected, along with five of the eleven people at two adjacent tables. Yet, none of the ten people at two other nearby tables became positive. The key factor here was airflow: The air conditioner blew air along the axis of the three tables with infections, then recirculated it. Diners at the tables off to the side were spared.[34]

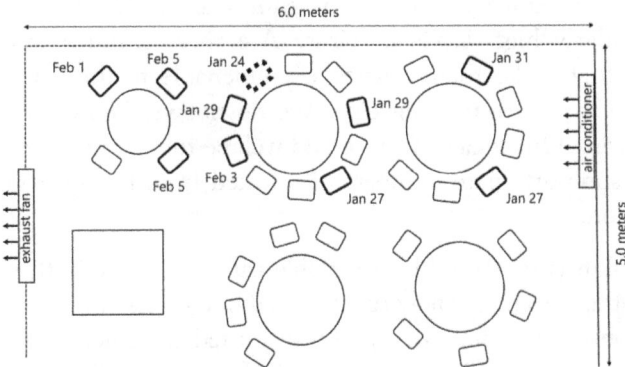

Figure 11: Sketch showing the arrangement of restaurant tables and air conditioning airflow at a site of an outbreak in Guangzhou, China, 2020. The dotted line shows the first index patient. The bold line shows the subsequently-infected individuals, with the dates they tested positive.

THE ROLE OF THE SUPERSPREADER

Evidence is increasing that the spread of the SARS-CoV-2 virus follows the Pareto Principle. Also called the "Law of the Vital Few," this law was determined when the Italian economist Vilfredo Pareto noticed in 1896 that 80% of the land in Italy was owned by only 20% of the population. This 80/20 pattern has been observed in many contexts.

Mathematical modeling of the transmission of COVID-19 suggests that 80% of cases are caused by only ~10% of the individuals infected.[35]

A Hong Kong study identified 349 local cases and traced over half of them (196) to six "superspreading events." The study succeeded in identifying one individual who appeared to infect 73 others. The researchers discovered that 80% of the local cases in Hong Kong were spread by only 20% of the individuals, with social gatherings being the key settings.[36]

In this Hong Kong study, the converse is also remarkable: A full 70% of the infected individuals did not pass the virus on to anyone.

In Boston, a superspreading incident at Biogen occurred early in the pandemic (late February of 2020) and was later analyzed, with tens of thousands of infections stemming from this single event. Remarkably, a full 3% of all of the cases in the United States and 1.7% of all cases worldwide came from this single business conference.[37]

A vital question remains unanswered: What are the specific characteristics of the 10% or 20% who infect so many others? Do these individuals have particularly high levels of virus? Are there features in their vocal physiology that make them particularly amenable to droplet and aerosol formation? Or is it something else? We do not yet know the answers to these questions. One area of focus could well be the nature of its dispersion, with the disproportionate numbers of infected from few individuals called overdispersion.

Overdispersion is measured by a variable called k, where both the original SARS and the current coronavirus is spread in clusters and super-spreading events, in contrast to influenza that is transmitted in a much more steady and linear fashion (measured by the variable r0). Thus the possibility exists that all the differences between counties, states, and countries can be accounted for not by non-pharmaceutical interventions nor health habits of a population, but by sheer chance.[38]

THE LIMITS OF THE CURRENT PARADIGM

The current testing paradigm is that of a steady increase of the number of real-time PCR tests being made available from the existing laboratory infrastructure. This means swabs, followed by RNA purification, then reverse transcription to convert viral RNA to DNA, and then real-time PCR for detection of viral RNA, with the result conveyed to the patient. Here we have hit the practical limits of the existing commercial landscape in the United States, which showed a lull of 800K tests per day over the Summer of 2020, but rose to over 900K tests per day in the early Fall.[39] This landscape is made up of a mixture of academic, government public health, and commercial laboratories of all sizes and capacities.

There are shortages of swabs, of the reagent the swabs are mixed into (called Viral Transport Medium), of the resins and chemicals needed for RNA purification, of the fluorescent dyes needed to label the short synthetic DNA strands (called primers), and of the primers themselves, in addition to the nucleotides and enzymes needed for the PCR process.

Essentially we have fallen into a trap: Test results are always too late to keep up with the spread of the virus. So, how can we fix this testing environment?

The X-Prize Foundation has been working on this issue. Over the Summer of 2020, they announced a contest to get Fast, Frequent, Cheap, and Easy (FFCE) diagnostic tests for coronavirus.

70 to 100 million tests are needed per month, which is tenfold higher than the current pace of testing. There are few existing alternatives to improve the current paradigm, although they are all moving forward.

An initial approach is to use the massive genetic throughput of next-generation sequencing, reviewed in Chapter 4, to replace the PCR amplification and real-time monitoring of fluorescence. Instead of instruments that can test a few hundred samples at a time, NGS can test tens or hundreds of thousands of samples at a time.

A second approach is to do away with nasopharyngeal and oropharyngeal swabs; these are swabs that go into your nostril or mouth to the very back and collect a patient's sample from these locations, as recent studies have shown that saliva is just as effective. One of the strange carry-overs from classical virology is that there is no standardization of the composition of the viral transport medium. Additionally, it has historically been optimized for the preservation of live, infectious virus and not for molecular testing.

A third approach is to enable multiple flu and other respiratory illnesses to test simultaneously (such as influenza, the four other coronaviruses that cause the common cold, rhinovirus, and other organisms with names like *Chlamydophila*). While there are point-of-care tests currently in use for the flu, having a multi-target test can untangle the confusion which can arise when determining whether a person has the flu, SARS-CoV-2, or one of the many versions of the common cold.

To deploy a saliva-based, very high throughput test targeting multiple viruses and bacteria, the existing NGS infrastructure would only require a dozen laboratories running at high (but achievable) volumes to serve the entire population of the United States. Instead of 800K or 1M daily tests we would have the capacity to perform 10M or more.

However, the logistics of what is known as the "last mile problem" would need to be worked out. The "last mile problem" refers to the problem communications companies face when installing fiber-optic cable for every household in a particular locale. Hooking up fiber to a small town or neighborhood is straightforward, but getting a fiber cable to cover that last mile to each individual home is an expensive and difficult challenge.

In the case of diagnostic testing, the "last mile problem" is how to get those samples from where they are collected to the central laboratory where they are processed. These are potentially infectious materials, which need special precautions and are also time-sensitive, as results need to be returned to the patient quickly.

Our centralized laboratories — such as the Laboratory Corporation of America (LabCorp) and Quest Diagnostics — have capacities to run 200,000 and 165,000 samples per day respectively (as of late August of 2020), representing almost half of the entire nation's testing capacity.[40] However, although they have sample transportation logistics infrastructure, they do not have the capability to enlarge their testing capacity tenfold.

Innovation will need to get us out of this paradigm and away from the central laboratory model to the point-of-need, which will be examined next.

A PERSPECTIVE ON SENSITIVITY

In the last chapter, we mentioned the at-home pregnancy test strip; a colorimetric test (a visible strip) indicating the presence of human chorionic gonadotropin hormone in the urine, which is a tell-tale sign of pregnancy.

The method of using antibodies to detect for the presence or absence of molecules is sensitive, and while that sensitivity is high (measured often in picograms per milliliter or one part per billion) its capability to detect virus molecules is nonetheless 100-fold less sensitive compared to a molecular PCR test. While a PCR-based test can detect less than 100 molecules of viral RNA, an antigen-based one can detect 10,000 molecules of viral protein.

This difference is important, as sensitivity directly impacts the false-negative rate (a less sensitive test may return a negative even though the virus was present, as it was just too low in its number of viral copies for the test to pick up). As discussed previously, false negatives (infected people being told they are virus-free) are an obvious threat. However, for this coronavirus pandemic, is identifying everyone with five copies of the virus truly necessary? The answer is no, as it stands to reason that it is the individuals with the highest amount of virus in the upper respiratory tract who would be the most contagious via aerosols or droplets from close speaking, coughing, or singing.

Additionally, even with five-molecule sensitivity, there was still a 20% to 38% false-negative rate with PCR, thanks to the inconsistent presence of the virus as referenced in Chapter 5, Figure 8.

Without fast, frequent, cheap, and easy testing, we are otherwise left to the existing methods of controlling infectious spread: mask-wearing, hand-washing, and avoiding close contact. A less sensitive test, but made widely available on an inexpensive basis, could be a trade-off worth bearing.

ONE GAME-CHANGING DEVELOPMENT FROM ABBOTT LABORATORIES

An antibody test detects virus protein, unlike the PCR tests that pick up virus RNA. The advantage of antibody tests is that they can be made inexpensively and depending on the implementation, they can be so simple to use that a dedicated CLIA-licensed laboratory is not required.

In the late Summer of 2020, Abbott Laboratories announced the development of a lateral flow immunoassay device (very similar to the aforementioned home pregnancy test). Simultaneously, they invested hundreds of millions of dollars to build manufacturing capacity while the device was under development. They announced not only the EUA approval of a 15-minute test that costs only $5, but also the manufacturing capacity to produce tens of millions of tests in September of 2020 and fifty million tests per month starting in October.

The construction of the paper card they have developed appears simple, but it is actually a fine example of materials engineering, biochemistry, and immunology. Inside, behind the paper card, is a material called nitrocellulose that wicks the sample across the material by capillary action.

Specific antibodies are picked up as the sample travels across the length of the nitrocellulose strip. Gold nanoparticle technology is used to indicate (by color) the presence or absence of a test signal. Positive and negative controls are built into the design.

If sufficient viral antigen is present, a colored line will appear in a test readout window; if absent, nothing will appear. A separate line must appear as the positive control line.

This technology has been used for decades with in-home pregnancy tests. Abbot calls their test for the detection of SARS-CoV-2 the Abbott BinaxNOW™ COVID-19 Ag Card. (The "Ag" in the name refers to Antigen, as it detects the virus protein by antibodies, rather than through virus RNA by PCR.)

Figure 12: The Abbott BinaxNOW card is a 15-minute, $5, CLIA-waived test based upon very similar technology to a pregnancy test strip.

Unlike other immunoassay antigen tests approved by the FDA under Emergency Use Authorization (there have been three others as of the Summer of 2020), the Abbott BinaxNOW technology has five features that clearly differentiate it from the rest: speed, convenience, cost, scale, and reporting.

In short, it is a 15-minute test available at a nurse's station in your local drugstore or urgent care clinic without the need for any instrument or device for readout; it will cost the patient $5, Abbott can manufacture 50 million per month, and Abbott has built a smartphone application to certify a negative result back to the patient.

Importantly, no instrumentation is needed at all, as the health professional simply reads the answer off the surface of the disposable card. Also importantly, this test is CLIA-waived, which means it can be used outside of what is called a high complexity laboratory under CLIA certification where existing PCR and other antigen tests are performed. These tests can go where the patients are, which means a number of very different contexts.

In essence, the instructions state: use the provided swab; take an anterior nasal sample (not from the nasopharyngeal area in the back of the pharynx, but from the front of the nose); apply six drops from the dropper bottle; insert the swab into the hole in the card and spin it three times; close the card with the adhesive strip; and read the results fifteen minutes later.

The accuracy of this test is of prime importance: At 97.1% sensitivity, this means a roughly 3% false-negative rate; at 98.5% specificity, this means a roughly 1.5% false-positive rate.

Under the EUA for the Abbott BinaxNOW test, this is for patients who have had up to seven days of symptoms of coronavirus infection. As you may remember from Chapter 5's Figure 8, the live virus (the bold line) disappears roughly one week after the start of symptoms.

One question arises: Could this test be used for screening the general population? What would be the public health implications? This is a topic of active discussion at the FDA and one to be examined next.

THE EUA TERMINOLOGY AND IMPLICATIONS OF SCREENING

While the FDA has given its authorization during the current emergency, this does not equate to FDA approval. Also, while CMS currently reimburses tests for COVID-19 that are authorized under the EUA, when the EUA is rescinded (and it is only a matter of time before the emergency subsides), the conventional requirement for FDA approval remains.

This means that the conventional approval pathway for diagnostics and all of its requirements still stand: the necessity for a pivotal trial, clinical samples, demonstration of vendor quality and vendor control, quality management systems, and interfering substance testing. The need for safe and effective diagnostics does not go away simply because we are in a pandemic; we need it even more so.

In our current regulatory system, the CDC handles disease surveillance; for example, influenza and choices for the upcoming seasonal flu vaccine. This occurs almost a year in advance due to the complexities of manufacturing the vaccine for the following year. The CDC also handles newborn screening for diseases such as those indicated by genetic testing for a faulty Cystic Fibrosis gene or a host of other conditions that can be treated if diagnosed early.

The FDA handles diagnostic testing approval, with all of its complexity and regulations for safety and effectiveness, as there are real consequences for inaccurate results.

As mentioned before, the Centers for Medicare and Medicaid Services (CMS) sets the higher standard of care as being "reasonable and necessary" over "safe and effective" and thus provides approval for payors.

Before considering how mass testing (i.e. "surveillance") for COVID-19 would work in absence of an Act of Congress, we have to address the basic feasibility and value of this approach to begin with.

For example, there are about 56 million school-aged children in the United States. To test one-fourth of them once per week would mean 56 million tests per month.

At a 98.5% specificity, there would be 840,000 positive results that would be false-positive ones, and let us also assume a 5% positivity rate. In this example, to confirm the positive test result by PCR would add to the overall cost. 56 million x $5 per Abbott test equals out to $280 million, and there would be 2.8 million positive test results (840K of them false-positives) all tested by PCR at $100 per test, so that means another $280 million.

You can see here that the cost of a $5 per person test becomes a $10 per person test, thanks to the need to re-test all potential positives. To save on the confirmatory testing, you would have to allow a full 30% of students (840 thousand out of 2.8 million, all of these false-positive students) to stay home for two weeks needlessly.

From an overall cost-to-benefit view, $280 million or $560 million may seem a small price to get back to in-person education. The larger question, however, is not so much economic as it is scientific: Would testing 25% of all school-aged children every week make a difference in the rate of transmission? There is reason to believe that the r0 (infection rate) could be effectively reduced to below 1 within only a matter of weeks if this testing went ahead.

However, this research, which would include the impact of this testing on the infection rate, would need to be completed and measured for such a mass screening effort to gain policy and economic justification and thus be implemented. Until then, the wearing of masks, washing of hands, and avoidance of close contact are the only tools we have.

CHAPTER 7: A REMARKABLE IMMUNE SYSTEM GONE HAYWIRE

Why don't ants catch the flu?

Because they have little anty bodies.

— **Cubmaster joke**

Person A, while at choir practice, had no idea she was going to be infected with SARS-CoV-2. Within a few weeks, several choir members she knew were diagnosed with COVID-19 and all of them recovered. It was almost a year after this, out of curiosity, that she took an antibody test at a clinic to see whether or not she had antibodies to SARS-CoV-2, and she was surprised by a positive result.

Person B was at a bar and brushed off any risk of infection. He was uncomfortable wearing a mask and hated what he viewed as an unjust intrusion on his personal liberties. He had been to a few places and talked with others without a mask before a mandatory order was put into place by the governor of his state. About three days after visiting the bar, he came down with the typical symptoms: a dry cough, a fever, body aches, and a loss of smell. His recovery was typical, requiring about two weeks of rest. He was irritated at all of the disruption in his daily routine, not to mention the two weeks of lost time at work.

Person C was scrupulously careful. She only went out around once per week and wore a mask, a face shield, and gloves at all times. She was 63 and only visited three places: her aging mother in a convalescent home (visits were not allowed for a long time at the beginning of the pandemic, but relaxed later); her 59-year-old brother who lived nearby; and a grocery store she went to only at opening when it was reserved for high-risk individuals. She had no idea how she was infected with SARS-CoV-2 given her limited scope of connections. While at home, she kept the pulse oximeter handy. Seeing the number dip below 90, she knew to head to the emergency room, and she was admitted to the hospital that morning. Her recovery was relatively quick as far as COVID-19 goes, and she credited her faith and her doctor that she did not need intensive care. The few stories her nurse was willing to tell her (from during the worst of the pandemic throughout the Northeast) made her shiver.

Person D was in his 40s and knew his risk factors were high. He was male and was aware that 60% of the hospitalizations and deaths were skewed against men; also, his BMI was over 30 and he had both diabetes and Chronic Obstructive Pulmonary Disease (COPD). Knowing all this, he took special precautions, but when he was infected with SARS-CoV-2 he had all of the classic symptoms (dry cough, fever, body aches, along with diarrhea) and was soon in Intensive Care on supplemental oxygen. His recovery was slow at 12 days; it took a little longer than the average ICU stay of U.S. survivors, which is 8 days. After a total of 18 days, he was finally discharged from the hospital. He knew that if he'd been put on a ventilator, his chances of recovery would have been less than 20% and was therefore relieved upon his recovery that he had been one of the lucky ones.

Person E was in terrific health, exercising religiously. In her 50s, she felt that the pandemic put a crimp on all of her running races scheduled months in advance, and like everyone else, she struggled to adapt to all of the different measures put into place by her local government. She knew her risk of contracting COVID-19 disease was extremely low outdoors, so she kept up her training at a lower frequency given that there were now no races to prepare for. She also didn't like wearing masks, even while grocery shopping or doing errands. When she got infected, she was sent to the emergency room. The same afternoon she was sent to the ICU, and she passed away a few days later.

These five cases of patients across the range of severity, from being asymptomatic (no symptoms at all) to dying of the disease, illustrate the insidious nature of COVID-19 disease. Over the course of human evolution our immune systems have kept us alive. The earliest recorded pandemic was in 450 BC in Athens at the height of the Peloponnesian War and is suspected to have been a typhoid fever that killed up to two-thirds of the population.

History is replete with periodic pandemics — notably the Black Death in 1350 (bubonic plague caused by a bacteria called *Yersinia pestis* carried by fleas), which caused so much death that England and France had to declare a truce during their war. During that time, the Vikings lost their ability to wage war against native populations and ceased exploration of North America and the British feudal system collapsed due to the rapid damage to their economy.

BACKGROUND OF SARS-COV-2

SARS-CoV-2 appeared in December in Wuhan, China and then spread rapidly, infecting 118,000 people across 114 countries in three months. This coronavirus has very similar molecular characteristics to the prior coronavirus outbreak SARS (Severe Acute Respiratory Syndrome) experienced in 2003. Before that outbreak, there had been a long history of coronaviruses, spanning all the way back to about the 13th to the 16th centuries, with the emergence of the common cold human coronavirus NL63 (abbreviated HCoV-N63).

There are four coronaviruses that cause the cold: strain NL63, strain 229E (around 1700 to 1800), strain OC43 (around 1890), and strain HKU1 (around 1950). The three recent ones are SARS-CoV (2002 to 2004), MERS-CoV (Middle East Respiratory Syndrome, 2012 to present), and SARS-CoV-2 (2019 to present). It is clear from the research that the origins of both SARS and MERS came with the jump from a bat host into an intermediate species (civet cats and camels respectively) and then a jump to infecting humans.

SARS infected eight thousand people and MERS infected twenty-five hundred, while SARS-CoV-2 has infected millions. The overall mortality for SARS was 774 (9.6%) and for MERS 858 (34%); for SARS-CoV-2, mortality is approaching a million dead (0.6%).

The two recent coronaviruses, unlike the four prior endemic coronaviruses that cause colds, attack the host (us humans) with severe symptoms. Thanks to research done on both SARS and MERS, several tools to combat the novel coronavirus were developed along the way, thus accelerating our ability to unlock some of the mysteries of how this virus evades the human immune system.

A LAYERED IMMUNE SYSTEM

Over millennia, our immune system has evolved into a wondrously complex and rich interlocking system with many intricate layers. This complexity, when the immune system malfunctions, is the root cause of chronic diseases such as rheumatoid arthritis, lupus, and Type I diabetes.

We have reviewed many of the advances over the past few decades in genetic and genomic data generation and analysis tools. These tools have been used to gain extensive knowledge about the intricacies of the immune system. Yet for biologists outside immunology, these complexities are too many to comprehend and the nuances too subtle to easily grasp.

Here, we do not intend to present a graduate-level seminar on immune-host defenses in infectious disease. The very idea of that brings back memories (which are not all positive) of my own senior year at the University of California, Los Angeles, and taking my last two required upper-division classes.

These seminars were replete with complicated mechanisms, drawings and names of cytokines, cell types, and molecular mechanisms of the invading organism (microbial or viral), as well as the immense armamentarium of the immune system.

However, with every year that passes, with new insights and new discoveries of novel molecules (chemokines were not taught yet when I was in school, for instance, only having been discovered in the late 1980s along with their receptors — something called G-Protein Coupled Receptors or GPCRs), one gets the feeling that we are on the edge of a vast ocean of details. Here in this section, we will give a broad overview of the immune system and how regulation and dysregulation of this system can help explain the large differences in COVID-19 disease as experienced between different individuals.

THE TWO TYPES OF IMMUNITY

The non-specific immune system is the first line of defense against infection. It is an ancient system, sharing elements in common with that of most animals, which works the same way across individuals. As the first line of defense, this innate protection is fast (literally working within minutes), allowing time for a specific response to be mounted.

For example, when you have a splinter in your finger, you will notice swelling, redness, and perhaps a build-up of yellowish or whitish fluid all within a few minutes. This is your innate immune system at work — a collection of white blood cells (with names like natural killer cells, mast cells, eosinophils...the list goes on) reproducing rapidly once kicked into action at the site of injury or infection and releasing chemical signals that amplify the message that a foreign invader is present.

Small fragments of the invader (whether SARS-CoV-2 inside the outer layer of your throat or airway lining or a bacteria on the surface of a splinter inside the outer layer of skin on your finger) are taken by specific cells to a nearby lymph node. Your tonsils, on either side of the back of your throat, are an example of a lymph node where these protein fragments (called antigens, from *anti*body *gen*erator) are presented by these cells to T-cells. The T-cells have literally billions of different types, all pre-programmed to interact with a single antigen and that one only for that single cell.

Of the practically infinite numbers of kinds of antigens our bodies can be faced with, the rearranged genes and the combinatorial math that can come out of that rearrangement will be examined next. The wide variation in response to the invading coronavirus is in no small part due to the intricacies of this system and shows how, at times, the immune system's failure can result in catastrophic consequences.

A PRE-EXISTING RECEPTOR PROTEIN

In the lymph node, a single T-cell with the pre-programmed receptor on its surface has been there all along, waiting for the presence of that particular antigen to come along and kick that cell into an activated, rapidly proliferating state where it will throw off many signals driving other immunological processes to occur. But as that receptor is a protein by nature, and since that protein had to be encoded by DNA, how did it get there in the first place?

There is some debate as to the overall complexity of the immune system and how many individual pre-programmed cells there are set up against invaders with foreign antigens. One estimate puts the number of different pre-programmed cells in an individual at around 11 billion, which is the number of estimated white blood cells that provide this antigen-recognizing function. Doing a theoretical calculation using the numbers of genes reshuffled (called Variable, Diversity, Joining, and Constant regions — or VDJ-C) shows that a much larger number is theoretically possible: ten to the nineteenth power.

It is difficult to visualize how large a number this is. There are about ten quintillion — or 10,000,000,000,000,000,000 — potential combinations of immune proteins that can be made against potential invaders. For those curious, that is 10 billion billion, the approximate number of cubic meters in all of the Earth's oceans, and the number of atoms in a single grain of salt.[41]

Antigens can be molecules never encountered before in nature; for example, a new lab-engineered chemical rearrangement of a protein has never been encountered naturally, yet humans can and do mount a defense against it. Similarly, SARS-CoV-2 had not been seen by any human before the initial outbreak in December of 2019, and for more than 99% of the people infected, an effective immune response is still activated and they recover.

Returning to the T-cell discussion, when a collection of white blood cells called (appropriately enough) Antigen Presenting Cells (APCs) present that specific antigen to the specific T-cell with the receptor to that antigen, the T-cell will then spark to life in a process called activation. That single cell with the specific receptor that recognizes a portion of the protein of the invader (in SARS-CoV-2 it is often the spike protein) will expand in number through cell division and become what is called a helper T-cell (Th) assisting other white blood cells in mounting a counter-attack. Other anti-spike protein T-cells become what is called a cytotoxic T-cell (CTC for short) meant to hunt and kill virus-infected cells. These T-cells also help B-cells that are preprogrammed against the spike protein antigen to multiply and produce antibodies.

The T-cell system works through cells (helper T-cells, cytotoxic T-cells, and memory T-cells), and the B-cell system works through the production of tiny proteins called antibodies. This overall system is referred to as the "two arms" of the immune system, working together in cohesion to mount a specific attack against an invading organism.

The innate immune system — the one that springs to life within minutes of a body getting a splinter, causing swelling and a rush of general immune activity — is the first line of defense. The adaptive immune system, which is specific to the invader and, in this discussion, set up against the SARS-CoV-2 spike protein, takes a few days to marshal its response.

After the specific T-cell gets activated, it multiplies and changes into several different types of cells. Each of the progeny cells inherits the specific spike antigen-recognizing capability. Different classes of T-helper and CTC cells derive from the original cell and interact with the B-cell system that produces antibodies. It takes around one to two weeks from the original infection for antibodies to be produced. The T-cells formed against the virus and B-cells that produce the anti-virus antibodies will have varying lengths of protection; immunity for smallpox confers immunity for life, having been shown to last as long as 88 years after vaccination. It is currently unknown how long immunity against SARS-CoV-2 will last.

In the case of SARS-CoV-2, antibodies stick to the spike protein of the virus, preventing its entry into new susceptible cells in the body. Cytotoxic T-cells sensitized to SARS-CoV-2-infected cells will work in cohesion with other immune cells to hunt and kill cells actively producing the virus.

It is this wonderfully tuned system gone haywire which is where the trouble lies. In a research study of 84 COVID-19 patients who had suffered severe respiratory failure within 8 days of the onset of symptoms, all of the severe cases had something called Macrophage Activation Syndrome (MAS): the disappearance of specific types of T-cells (called CD4+ and Natural Killer cells, a type of cytotoxic T-cell).[42] Macrophage Activation Syndrome has historically been associated with chronic arthritis in young people and is a potentially life-threatening condition. Arthritis, as you may know, is also a disorder of the immune system.

In cases of severe COVID-19, the pneumonia-like fluid build-up in the lungs leads to respiratory failure. Other effects of severe disease include occasional general organ failure and nerve damage, including loss of the sense of smell.[43] All of these effects are caused by a malfunctioning immune system; this is currently an active area of study.

THERAPIES TIED TO THE IMMUNE SYSTEM

In the early stages of the pandemic, it was observed in late-stage COVID-19 patients that a massive inflammatory response (called a "Cytokine Storm" due to the increase of immune system cytokines such as IL-6 and interleukins and tumor necrosis factors) could be treated with medications intended to reduce their presence and effect. Several treatments were explored to blunt the effect of a cytokine storm, such as anti-IL-6 monoclonal antibody therapy, as well as the use of corticosteroids such as dexamethasone and the use of an anti-cancer immunotherapy agent PD-1 checkpoint inhibitor (due to the role the immune system plays in cancer). An external device that absorbs cytokines or even dialysis have been proposed as treatments[44]. However, even the definition of a cytokine storm has come into question.[45]

Immune-system modulators (either turning the system up or down) have been used with sporadic effectiveness in past respiratory viral outbreaks; using these and other clues, there are currently over 700 unique active compounds in development to treat COVID-19 and several hundred clinical trials underway. The industry organization, Biotechnology Industry Organization, maintains a COVID-19 Therapeutic Development tracker[46] and, to dig further into the details of the latest treatment guidelines, the U.S. National Institutes of Health (NIH) maintains COVID-19 Treatment Guidelines with the latest official recommendations for anti-viral therapy, immune-based therapy, adjunctive therapy, and concomitant medications.[47]

Of the immune therapies mentioned above, one promising treatment will be examined next: the corticosteroid dexamethasone.

Dexamethasone is a synthetic corticosteroid first approved as a treatment for rheumatoid arthritis in 1958. Further study and understanding of its applications, as well as the availability of generic forms of the drug, meaning that approximately 1 million prescriptions for dexamethasone are issued in the United States each year.

Dexamethasone had been tried during the prior SARS and MERS outbreaks, but the data there was not significant enough to offer any firm conclusions. An international, random, case-controlled clinical trial in June of 2020 demonstrated that this widely available steroid reduced deaths in severe COVID-19 cases by a full one-third. A U.K. physician called it a "startling result…" and remarked that "it will have a massive global impact."[48]

During dexamethasone treatment the doses are small, and no adverse side effects were observed for any of the treated patients. It is important to here make a distinction between the two ways clinical science is advanced: the random controlled trial versus observational (also called retrospective) analysis.

RANDOMIZED CONTROLLED TRIALS VERSUS RETROSPECTIVE ANALYSIS

Randomized Controlled Trials are considered the "gold standard" of evidence to show the effectiveness of a treatment or procedure. A placebo is the control group; the untreated individuals. A randomly selected group is the trial group; the one that gets the treatment.

There can be what are called "arms" of a trial; one trial arm could be a group that gets one dose of a medication, a second trial arm could receive a different dose, and a third could receive two doses spread apart.

In case you were wondering, a "double-blind" trial is one where both the person receiving the treatment (or placebo) and the physician or nurse administering the treatment do not know what is being administered. The opposite is "open-label," where everyone knows what a patient is receiving.

The alternative kind of evidence gathering is the retrospective or observational study. This is where you go back into the medical records and collect data from individuals with a certain condition who were treated in a certain way and then compare them to other individuals with the same condition who were treated differently.

The problem with an observational study is one of bias. Medicine is both a science and an art: There are standards for treatment; however, a large amount of latitude and discretion is given to the physician where there are a multitude of treatment options or where there is an emerging disease such as COVID-19. It is difficult to control for all of the subtle signals and characteristics of different patients that may affect results, so these treatments were not random, as in the RCT.

I must offer a word about "statistical significance": Anecdotes are not data, confusion between correlation and causation happens all the time, and our own bias toward making hypotheses and working models leads us in many different directions. The existing result with remdesivir has been called a "proof of concept"—that there is some anti-viral activity that speeds recovery. Importantly, no significant difference in fatality rate from COVID-19 was observed in its initial study, but more time and larger studies are needed. This work is ongoing.

UPCOMING ANTIBODY-BASED THERAPIES FOR COVID-19

One powerful approach is to tackle the virus relatively early on in the course of COVID-19 with neutralizing antibodies. The term "neutralizing" is used because all antibodies are not equivalent; these are antibodies that prevent the live virus from propagating, which may well be different in effect than an antibody that binds to a virus component. For example, you can have an antibody against a component of the interior to the virus that has no effect on the virus's propagation throughout the body.

Instead of manipulating the infected host's (the patient's) immune response to work correctly, you simply aid it with antibodies produced elsewhere, whether from a recovered patient or from a pharmaceutical company.

One therapy using convalescent plasma from recovered COVID-19 patients has been approved under EUA. However, this approval was based upon retrospective observational data, not a prospective randomized controlled trial.

Per the NIH's technical guidance document, this treatment authorization has several unusual recommendations in it, as follows:

There are insufficient data to recommend either for or against the use of convalescent plasma for the treatment of COVID-19.

Convalescent plasma should not be considered standard of care for the treatment of patients with COVID-19.

Prospective, well-controlled, adequately powered randomized trials are needed to determine whether convalescent plasma is effective and safe for the treatment of COVID-19. Members of the public and health care providers are encouraged to participate in these prospective clinical trials.[49]

While authorized for emergency use, this encouragement to participate in a prospective trial is ineffective. As a practical matter, patients will insist on getting the treatment rather than risk being placed in the control (placebo) group. We may never know the usefulness of convalescent plasma therapy, although one prospective trial out of Vanderbilt University is going to attempt it nonetheless.[50]

A FLOOD OF MONOCLONAL ANTIBODIES

Another promising approach for therapy is to develop monoclonal antibodies against the coronavirus in particular or to specifically modulate the body's immune response with an antibody. Candidates from Eli Lilly, Regeneron, and several others are currently under active development, and a review publication lists 46 candidates in preclinical through Phase III randomized controlled clinical trials.[51] As you remember, monoclonal antibodies are identical and specific against a particular antigen and would be most effective early in the course of the disease. Notably, the U.S. President was treated with a monoclonal antibody combination called REGN10933+REGN10987 from Regeneron in October of 2020 under compassionate use clearance by the FDA.[52] (The FDA term "Compassionate Use" is a granting of access to a drug in a clinical trial if all other options have failed and the patient might die without the drug.)

THE CHALLENGE OF PRODUCTION SCALE

As you remember, antibodies are tiny proteins which are 1/1000th the size of a typical immune cell. The developers of the therapy started with blood donations from COVID-19 patients to isolate those special immune cells that have the specific rearranged genes which produce antibodies tailored against SARS-CoV-2 infection. Going all the way from a cloned gene in R&D through to a 10,000-liter bioreactor requires an enormous amount of work in a compressed timeline.

Although the processes for manufacturing small molecule drugs are fundamentally different, the challenge of scale becomes the same.

The plants needed to produce monoclonal antibodies or small molecules have strict regulations because they are producing medicines to be injected into people for life-saving therapy. Known as cGMP ("current Good Manufacturing Practice"), the standards are maintained by the FDA in a fashion that describes the specific meaning and enforcement for different levels of manufacturing separate kinds of products for human use. In their database, they currently maintain over 50 documents specific to cGMP.[53]

Practically, this care means that a whole infrastructure is needed: people, processes, and information with specialized skills such as quality assurance and quality control. Each incoming item as part of the supply chain has to have parameters for testing and records of how that particular manufacturing batch met the set requirements. You are required to have a documented method for what to do if one component does not match your own requirements. That means intermediate testing of your internal manufacturing process, whether for contamination by unwanted intermediate chemicals used in processing or for the proper specifications for your product. In addition to this, bacterial contamination is a possibility. Internal quality checking is a way to reduce the risk of losing literally millions of dollars in investment in all of the components that comprise the production of any medication used for injection.

There are standards of cleanliness which the USDA maintains for food production (the number of bacteria per milliliter your milk can contain for example). For any medicine to be injected into the body, we are talking about a whole different level of purity.

And at its conclusion, there must be an evaluation of the final product before it ships. Does the medication do what you as the researcher said it was supposed to do? There has to be a functional test of whether or not the monoclonal antibody or drug can block entry of the SARS Coronavirus-2 into susceptible cells, as well as all of the other tests required to ensure it is free from both toxic and bacterial contaminants.

These monoclonal therapies are not only evaluated for preventing death from COVID-19 as a therapy; they are also being evaluated for preventing infection or COVID-19 disease in household contacts.[54]

One immunosuppressant, hydroxychloroquine, as a treatment for COVID-19 will be discussed next.

AN EARLY TREATMENT BECOMES POLITICAL

The antimalarial drug hydroxychloroquine (HCQ) held promise at the beginning of the pandemic and then became a political football. Previously, this drug was primarily taken as an effective treatment for chronic sufferers of the immune disease systemic lupus erythematosus. For lupus patients, HCQ had been demonstrated to increase survival and reduce kidney and other major organ damage, in addition to protecting cardiac health.

HCQ was shown in laboratory cell cultures to be effective against both SARS-CoV and SARS-CoV-2. The FDA gave an Emergency Use Authorization (an EUA) for hydroxychloroquine use in March as the epidemic in the United States took hold.

HCQ was later shown not to help in either post-exposure prevention or late-stage COVID-19 disease. The FDA revoked the EUA for this particular use a few months later in June.

For post-exposure prevention, a randomized, double-blind, placebo-controlled trial was undertaken where household members or front-line healthcare workers were exposed to someone with confirmed COVID-19. High-risk exposure was defined as exposure lasting at least 10 minutes and occurring within 6 feet of distance without either eye or mask protection; moderate risk was such exposure without eye protection. Four days after exposure, trial participants were either given multiple doses of HCQ or a placebo (and with the study being "double-blind," neither the physician nor the nurse administering the medication nor the patient knew whether they were receiving HCQ or a placebo).

Of 821 participants, a total of 107 individuals became infected, with 49 of those being in the HCQ group and 58 in the placebo group. This difference was not statistically significant, with the HCQ group observing a larger number of side effects (40% versus 17% between the test and placebo groups).[55]

Other studies had come out previously to rule out HCQ for late-stage disease; over fifty studies have now been conducted as of October of 2020, and they have not shown significant effectiveness in decreasing the mortality or severity of COVID-19. However, with the politicization of treatment, individual physicians and others have compiled anecdotes and observations. Also, studies have been done analyzing retrospective data.

A prominent retrospective study from the Henry Ford Clinic was picked up by politicians to show a 50% lower death rate with HCQ treatment[56].

This retrospective study, however, has a large, confounding element: the use of steroids (in particular dexamethasone, which was shown to be highly effective). By combining both HCQ and the use of steroids, this headline statement of "50% lower death" fails to separate the effect of steroid treatment.

To repeat: this study looked backward after the results were known in their medical records, subject to the limitations of bias in terms of who was selected to be treated versus getting no such treatment at all.

As previously stated, what is called "Standard of Care" medical treatment comes after years of prospective randomized controlled trials, which are painstakingly and carefully performed, and not through observational, retrospective ones.

As an example of a retrospective study that did not hold up to a randomized controlled trial, take the relatively common affliction of arthritic knees. You would think that knee surgery could be effective in helping arthritis; this was an observational, retrospective study showing promise. Once a randomized controlled trial was undertaken, however, the effect of the surgery was demonstrated to have no benefit over existing treatment.

For another example, it was thought that estrogen replacement therapy for postmenopausal women would help prevent heart attacks. Retrospective studies showed a powerful effect. Again, several randomized trials showed conclusively that hormone replacement was of no benefit.

For hydroxychloroquine, one large randomized controlled trial by the U.S. NIH (called the ORCHID Study) was halted before it could be finished, citing the side-effects of patients in the trial.[57] A few weeks later, the WHO (in its own trial called Solidarity) stopped a trial before it could be completed, citing the ineffectiveness of HCQ.[58] What these two trials used was high-dose HCQ, on the order of 4 times the amount (9,400 mg over five days compared to 2,400 mg) normally used for chronic conditions such as lupus with minimal side-effects.

One recent retrospective study looked at the use of low-dose hydroxychloroquine in early-stage COVID-19 patients for the entire country of Belgium. These 8,075 patients were diagnosed before May 1, 2020, and discharged by May 24, 2020, and while it was not an RCT, the large number of patients enabled a fair comparison between untreated and treated individuals.

The results of the retrospective Belgium study were impressive: Those who did not receive HCQ had a death rate of 27.1%, while those who received HCQ alone (with no other combination of medications like the commonly referenced antibiotic azithromycin and the mineral zinc) had a death rate of 17.7%.[59] This is a reduction in the rate of death by a surprising 34%.

Nonetheless, other work with the gold-standard RCT showed no effect at a lower dose in mild and moderate cases of COVID-19.[60] Thus, after all of the effort, publicity, and controversy, the current NIH treatment recommendation weighs in against hydroxychloroquine with or without the antibiotic azithromycin.[61]

To conclude this section on the immune system, we will take a look at those who are suffering from the long-term effects of COVID-19.

THE CYTOKINE STORM AND LONG HAULERS

With an over-reactive immune system triggered while fighting uncontrolled SARS-CoV-2 virus, replication in cases of severe COVID-19 produces the aforementioned cytokine storm. Many cytokines (with an acronym soup like IFN-α, IL-6, TNF-β) interact in a non-linear and complex fashion during the course of the disease, leaving long-lasting side-effects that are increasingly being acknowledged.

The reports of COVID-19 symptoms lasting for months have led to patients being called "long-termers" or "long-haulers" who have formed support groups online. One group organizing over the communication app "Slack" has over 14,000 members[43]. The most common symptoms are extreme fatigue, muscle weakness, low-grade fever, an inability to concentrate, memory lapses, sleep difficulties, and shortness of breath, extending beyond 12 weeks from the beginnings of first symptoms.[62] Additional symptoms include headache, diarrhea, loss of taste and smell, skin rash, and chest pain resulting in emotional, physical, and financial stress. Experience with SARS and MERS indicates a small percentage of patients will take years to fully recover. In the case of SARS, one study followed 71 patients for 15 years, and more than a third of those patients had reduced lung capacity.[63]

The list of additional possible long-term consequences looms large for these patients. Not only is lung tissue susceptible to coronavirus infection, but any tissue that has the *ACE2* receptor on it, including the endothelial cells that line blood vessels, is also susceptible. Thus, any organ can be affected by the damaged oxygen supply and inflammation that infection brings, such as lung scarring; blood clots in major organs (causing stroke if in the brain, pulmonary embolisms in the lung, and kidney failure, in addition to deep vein thrombosis); heart damage; and neurological and cognitive damage.[64]

Critically ill children have had what is described as Pediatric Multisystem Inflammatory Syndrome (PMIS), a severe immune disorder that shares similarities with Kawasaki Disease. In these patients, high fevers, severe abdominal pain, and abnormal immune markers of inflammation are observed.

Long term COVID-19 is certainly in the minority of cases, but it is increasingly being recognized. These symptoms and effective treatments for them will be investigated and pursued for years to come.

Yet this section on the complexity of the immune system's dysfunction will conclude on an optimistic note — the potential for underlying immunity due to prior exposure to other coronaviruses such as the common cold.

DIRECT EVIDENCE FOR IMMUNITY

One natural experiment was reported recently in Seattle, Washington, where all 122 of the crew members of a fishing boat were tested both for the presence of active virus using PCR as well as for the presence of antibodies against SARS-CoV-2. All tested negative by PCR; however, three not only tested positive for the anti-SARS-CoV-2 antibodies, but the scientists proved it had neutralizing activity (that is, it prevented entry into living susceptible cells in laboratory culture). Three others had a low level of anti-SARS-CoV-2 antibodies (near the lower limit of detection cutoff for the antibody test they used, an automated immunoassay system from Abbott Laboratories called the Abbott Architect).

The crew somehow got infected with SARS-CoV-2 (the research paper does not discuss how the virus was able to spread among a group all testing negative), and after 18 days at sea, they had to return when one of the crewmembers was so sick that they needed hospitalization. Tested for both virusese via PCR and antibodies using a serology test and followed up on for over a month after their return, a full 85% of the crew (105 out of 122) were infected.

Significantly, none of the three crew-members who had high neutralizing antibodies before boarding the fishing vessel got infected. This is the first direct proof that humans with neutralizing antibodies (and the T-cells and B-cells specific against the virus, which is much more difficult to test for) has been conferred with protection against active infection. This will become important later in the discussion on vaccines, and this direct information gives us clues as to how effective that protection can be.

Your immune system is a wonder of nature; an area of intense research whose secrets are revealed only begrudgingly. With layers and layers of complexity, immunology is one of the areas of biology that is demanding, vitally important, and completely responsible for the wide variation in outcomes of COVID-19 disease.

CHAPTER 8: ADJUSTING STANDARD OF CARE TREATMENT IN REAL-TIME

Blessed are the flexible, for they shall not be bent out of shape.

—**Betsy Shirley, Buck Brannaman's foster mother in The Horse Whisperer**

The news came somewhat as a shock to those familiar with medical device manufacturing: Ford Motor Company and GE Healthcare, as well as General Motors with Ventec Life Systems, would be combining their manufacturing technology and skilled labor to manufacture and assemble tens of thousands of ventilators. It would prove to be unnecessary, but in early April of 2020 with the pandemic in its early phase, having sufficient numbers of these complex devices was seen as a linchpin to ensuring the fewest number of deaths possible.

Those familiar with the exacting standards of medical device manufacturing would shudder at the thought of assembling these fragile and exacting devices in an automobile manufacturing environment. Yet the need for expanding capacity was there, and creative solutions were found.

Other grassroots efforts focused on the lack of Personal Protective Equipment (PPE), and instructions on how to adapt 3D-printing or use office supplies were eagerly shared over social media.[65] Other groups sprang up to sew homemade masks, including MasksNow as they refined designs and organized ad-hoc sewing groups nationwide to reduce the demand by consumers for hospital-grade PPE.[66] At one point relatively early in the pandemic, there was a popular Facebook group crowdsourcing a new ventilator design.[67]

Companies, groups of concerned citizens, and other volunteer groups all wanted to do what they could to help reduce the spread of the virus, the burden on the healthcare system, and fatalities. This responsiveness to need at the national and local level was palpable; people limited in-person gatherings and employment on many levels, as they wanted to contribute and make a difference.

In the fearful and uncertain early phases of this pandemic, there was no time to waste as lives lay in the balance.

The ongoing adjustments to new information, the exploitation of pre-existing systems in new ways, and creative methods of adaptation to the realities of a COVID-19 pandemic will be examined here in the context of treatment by healthcare workers.

ROBUST COMMUNICATIONS IN AN EPIDEMIC

One of the remarkable capabilities on display throughout the COVID pandemic in the United States has been the robustness of the communications infrastructure. Telephone call volumes exceeded 800 million wireless calls per day — twice that of the highest volume on Mother's Day (which historically shows the highest volume throughout the year). Additionally, and perhaps more importantly, the internet infrastructure has not buckled under the immense increase in volume compared to pre-pandemic volume.

This increase in communications volume — an estimated 25% of overall internet data volume where so much of society and commerce has moved online — also enabled worldwide communication between front-line physicians, nurses, and other medical staff in far-flung locations (for example, from Italian emergency room staff communicating to the U.S.) in private forums, individual group chats, and distributed emails. Many thoughtful ideas were attempted in these communications, a few observations were made, and enough people communicated their findings in academic preprints and formal journal submissions that clinical trials were started from these communications.

One important aspect of this communication was publishing in academic journals. Scientists would submit their article to a paid publisher along with a fee for publishing their research. A multi-billion dollar publishing industry depends on author submission fees; advertising and subscription fees for arranging for peer reviewers (who volunteer their expertise); and other various necessary services, including editing, formatting, publishing in print and digital, and distribution.

The importance of these articles upon the reputation and relative significance of scientists' work cannot be overstated. The aphorism "publish or perish" is well-known among scientists; the name of a journal indicates relative prestige, as the journals *Science* and *Nature* take the top spots, along with related journals such as *Science Translational Medicine* and *Nature Nanotechnology*, depending on a scientist's specific sub-specialty; other prominent biology journals include *PNAS* (*Proceedings of the National Academy of Sciences*) and *Cell.*

In the medical field, two top journals are *The New England Journal of Medicine* (NEJM) and *The Journal of the American Medical Association* (JAMA). Additional journals of importance to physicians in this crisis have included *The Lancet* and the CDC's own *Emerging and Infectious Diseases Journal.*

THE USEFULNESS OF A PRE-PRINT

A few years before the pandemic, a movement began called "Open Access" that used peer reviewers to ensure a standard of quality of research in addition to an open availability model that eliminated both the profit motive and onerous subscription fees for academic institutions (universities and research institutes), companies, and individuals who would subscribe to these publications. In a parallel effort, pre-print servers were established to make available manuscripts before they went through peer review (as well as editorial changes and output into a polished format).

This mechanism to make research available is invaluable. Before the advent of open access, current research would first be presented in a specialized conference, with the results in the hands of a traditional publisher that would take some weeks or months to publish it in its final format. Often that time is needed because of the request for additional experiments or analyses; during this process, almost half of research journal submissions are rejected, starting the process all over again.

Through the aforementioned high-speed communications networks, findings in one corner of the world could be rapidly and efficiently communicated to other specialists who needed the relevant information. Even though these research results were not peer-reviewed, which means they had not gone through rigorous screening by other experts in their field who had not been directly involved in their research, the advantage of sharing these results quickly meant that the results could be applied or other research directions started. Although pre-prints were encouraged during the Ebola and MERS epidemics, their use was limited.[68]

Between January 1st and April 30th of 2020, over 16,000 journal articles were published about COVID-19, with over 40% of them coming via pre-prints. These breaking research articles have been picked up by major media outlets in addition to social media platforms, receiving an "unprecedented amount of attention from scientists, news organizations and the general public, representing a departure for how preprints are normally shared."[69]

Examples of this fast spread of innovation will be discussed below; from individual communication to formal research to pre-print dissemination of information affecting how COVID-19 has been treated.

FAST INNOVATION IN TREATMENT

Two advances in the treatment of COVID-19 implemented high oxygen cannulas and placed severe COVID-19 patients in a prone (on the stomach) lying-down position. These two changes became standards for treatment just as the news about the relative ineffectiveness of ventilation was demonstrated: Only 12% of COVID-19 patients on a ventilator survived in a study of 5,700 hospitalized patients in New York.[70] To put this number in context, patients with other types of respiratory distress on a mechanical ventilator have a survival rate of about 80%. Early optimism over the usefulness of ventilators (and the publicized effort to increase their manufacture) proved premature.

Having severely low oxygen levels in the blood is called hypoxia, and this is one of the hallmarks of severe COVID-19 disease; however, it turns out that low oxygen levels in a COVID-19 patient are not nearly as life-threatening as the same low oxygen levels in a sepsis or ordinary pneumonia patient. In a medical paradox, the so-called "silent hypoxics" among COVID-19 patients feel and speak normally and in general show no signs of respiratory distress despite abnormally low oxygen levels.

Face masks used for sleep apnea (called CPAP, which stands for Continuous Positive Airway Pressure) or for congestive heart failure (called BiPAP for Bi-Phasic Positive Airway Pressure) have been used in addition to nose prongs (cannulas) to deliver large amounts of oxygen to COVID-19 patients. It is worth noting that with these tools, the risk of producing contaminating aerosols with an invasive ventilator is avoided, whereas a tube needs to be inserted with a ventilator's use.

ACCELERATING RESEARCH AND PRACTICE

In ordinary times, the practice of medicine changes slowly. Randomized case-controlled studies with placebo and trial groups require time and resources, and official societies and committees make recommendations based upon a slowly building body of evidence that typically takes several years to accumulate.

In the middle of a pandemic, with cases accelerating exponentially, there is no such luxury.

National healthcare with accompanying interoperable electronic medical records offer vast resources for discovery. In 2009, then-President Obama signed the American Recovery and Reinvestment Act, with $19 billion set aside for Health Information Technology for infrastructure and incentives to adopt electronic health records. Over ten years later, the United States still suffers from a Balkanized system with over 300 different EHR (Electronic Health Record) systems in use.

OpenSafely is a huge UK National Health Service (NHS) study and resource, encompassing a full 40% of the entire population of England (some 17.2 million adults). One study analyzed all primary care records associated with 10,926 COVID-19 deaths.[71] By looking at specific behaviors, as well as characteristics and pre-existing conditions, the risk of death from COVID-19 could be exquisitely calculated and reported in the form of a Hazard Ratio [HR]. (Here, for the sake of simplicity, we will not include the 95% confidence interval, which is the statistical measure of the reliability of an estimate.)

A hazard ratio of 1.0 means there were no differences at all between a treatment group and a control group. In this case, it is the relative risk of death we are talking about; thus, if you are male, this study reported the hazard ratio of males over females to be 1.59. Translated, that means as a male you have a 59% greater risk of death based on your gender if you have COVID-19.

From the OpenSafely report, the researchers set the age of 50 to 59 as the reference point, where the hazard ratio is 1. For those in the age range of 40-49, the hazard ratio is 0.25; that is, you are four times as likely not to die of COVID-19 if you are in your 40s compared to if you are in your 50s. If you are in your 60s, the HR is 2, meaning you are twice as likely to die as someone in their 50s. If you are in your 70s, the hazard ratio spikes up to 5. If you are in your 80s, it is something like 15.

Importantly, by ethnicity, the reference hazard ratio was set at 1 for being white; an elevated HR of 1.48 and 1.45 for black and South Asian ethnicities respectively were observed (a 48% and 45% greater risk of death for these populations). Naturally, many other accompanying medical conditions (called comorbidities) increased the risk of death, with the following at an HR of 2: diabetes, obesity, cancer diagnosis within the past 12 months, chronic respiratory or kidney disease, and stroke. Notably, conditions with an HR of 3 or 4 include hematological (blood) cancers, reduced kidney function, and organ transplant.

One of the most surprising results is a reduced risk (slight, but statistically significant) of a current smoker status. The "never smoker" group was the reference group — an HR of 1 — and former smokers have an elevated risk of 1.3 (a 30% greater risk of death by COVID-19). Current smokers are 5-10% *less* likely to die of COVID-19 compared to people who have never smoked. Remember, this is just an interesting observation, and not a recommendation to start smoking!

Naturally, with a single standardized dataset, an examination of treatment differences can also be done in retrospect, but the problem of bias remains. There are so many factors involved with individual differences overlaid with all of the treatment options, the data cannot be compiled and analyzed fast enough to tease out cause-and-effect relationships for differences in treatment using Electronic Health Record (EHR) datasets.

In a time where Big Data is currently analyzed by businesses for insights and trends to increase purchases or engagement (for example, if you are shopping at Amazon, you get suggestions to buy things based upon your purchase history; while chatting on Facebook, you get suggestions on who else to send a friend invitation out to), the personal healthcare data held by hospitals and institutions is a set of data that promises to yield tremendous insights in the future.

One prominent example in the recent past was the IBM effort called Watson Health intended for personalizing cancer medicine. This was a $5B effort and a notable commercial failure with weak results and few studies, along with a management and financial scandal at the top medical center, M.D. Anderson Cancer Center in Houston, Texas. Other companies are continuing this big data and machine learning effort in healthcare, including Nvidia, Tempus, and Syapse.

The tantalizing potential exists to use EHR "big data" and several attempts at doing so for COVID-19 are being made.

ROOM FOR ALTERNATIVE MEDICINE

During a pandemic, the science of medicine gives way to the art of medicine. Following the Hippocratic tradition, the first principle of "first do no harm" applies.

In Norfolk, Virginia, there is a small medical school called the Eastern Virginia Medical University where a South African-trained internist named Dr. Paul Marik has developed a 30-page treatment protocol for COVID-19 (called the "EVMS Critical Care COVID-19 Protocol") and a two-page summary ("Marik COVID-19 Protocol Summary")[72]. What is unusual about the Marik Protocol is its use of supplements for mild cases of COVID-19, as well as recommendations for potential prophylaxis (i.e. prevention of disease).

The list of supplements with anti-viral properties has some published evidence to support it. (One of the reasons for the length of the protocol is the inclusion of over 200 references spanning 10 pages.) Melatonin and Zinc were early recommendations and work on Vitamin D (direct correlation of low Vitamin D to COVID-19 mortality has been demonstrated) led them to add it to the list.[73]

Recommendations for hospitalized patients with mild symptoms (and suggested as optional for patients at home) interestingly include famotidine, popularly known as Pepcid®, which works by reducing the amount of acid the stomach produces by blocking a molecule called histamine-2. Pepcid is used to treat ulcers and acid reflux and several prospective trials are underway.

In an observational study, Pepcid administered during hospitalization had a hazard ratio for death or intubation (being put on a ventilator) of 0.42; that is, there was a 58% greater chance of dying or intubation without the famotidine.[74]

Certainly, the addition of a common, over-the-counter antacid medication with a history of many decades of safe usage would do no harm for those who would take it as a preventative measure. At one point, there was a recommendation for HCQ. However, the list no longer includes it, and now contains a paragraph with the evidence for its controversial nature, its current exclusion from their recommendations, and the need for staying informed on the evidence.

I am not a physician, so I am sharing what the Marik Protocol states for prevention and at-home symptomatic COVID-19 patients as an informative resource. Understand that the nature of their recommendations will change. The following is for informational purposes only; please note that BID is a physician Latin acronym for "twice daily," and ASA is shorthand for aspirin.

From the two-page Critical Care COVID-19 Management Protocol document dated September 2, 2020[72]:

Prophylaxis

While there is very limited data (and none specific for COVID-19), the following "cocktail" may have a role in the prevention/mitigation of COVID-19 disease.

- Vitamin C 500mg BID and Quercetin 200-500mg BID

- Zinc 75-100mg/day (elemental zinc). Zinc lozenges are preferred. After 1 month, reduce the dose to 30-50mg/day

- Melatonin (slow-release); begin with 0.3mg and increase as tolerated to 2mg at night

- Vitamin D3 1000-4000 I.U./day

- Optional: Famotidine 20-40 mg/day

For those "Mildly Symptomatic patients (at home)", the Malik Protocol has the following recommendations. Again, the same caveats apply as above.

- Vitamin C 500mg BID and Quercetin 200-500mg BID

- Zinc 75-100mg/day (elemental zinc)

- Melatonin 6-12 mg at night (the optimal dose is unknown)

- Vitamin D3 2000-4000 I.U./day

- Optional: Ivermectin 150-200 ug/kg (single dose)

- Optional: ASA 81-325 mg/day

- Optional: Famotidine 20-40 mg/day

- In symptomatic patients, monitoring with home pulse oximetry is recommended.

- Ambulatory desaturation below 94% should prompt hospital admission[72].

Ivermectin is an FDA-approved drug to treat specific internal parasitic worms in humans and external headlice; it is also used for preventing heartworm in animals. The FDA has specifically stated that it should not be used to prevent or treat COVID-19.

From the FDA webpage:[75]

> *While there are approved uses for ivermectin in people and animals, it is not approved for the prevention or treatment of COVID-19. You should not take any medicine to treat or prevent COVID-19 unless it has been prescribed to you by your health care provider and acquired from a legitimate source.*

This chapter will conclude with a new antiviral therapy approved under EUA for the treatment of COVID-19 named Remdesivir.

THE ANTIVIRAL DRUG REMDESIVIR

Of several repurposed and new antiviral drugs, Remdesivir was originally developed by Gilead Sciences as an antiviral, small-molecule drug against Ebola, designed to interfere with the RNA replication of the virus. It is the only one to date that has shown benefit from a prospective RCT, although the data so far has only shown reduced time in the hospital (from 15 days to 11 days in patients with serious markers of COVID-19 disease). A reduction in the death rate from COVID-19 has not been demonstrated so far.[76]

Different antiviral therapies are going through prospective randomized clinical trials. An anti-flu medication approved for flu in Japan is one (called Favipiravir, or brand name Avigan® produced by FujiFilm Holdings), a Merck compound under development called MK-4482 is going through a large Phase III trial, and the aforementioned anti-parasitic drug called Ivermectin (as part of the Marik Protocol) has a large Phase III trial underway in Singapore. Progress on these and other therapies can be tracked online.[77]

For clues on where research is headed, we will next look at natural experiments related to SARS Coronavirus 2 infection upon human genetic variation.

CHAPTER 9: HETEROGENEITY, VARIATION, AND NATURAL EXPERIMENTS

The headline was dramatic. "Healthy 30-year-old Teacher Dies Suddenly from Coronavirus." The CNN video story from April 2, 2020, continues, "Ben Luderer, a 30-year-old New Jersey teacher and coach, passed away only days after contracting coronavirus. Brandy Luderer, his wife, shares the heartbreaking details with CNN's Alisyn Camerota."[78]

It is hard to imagine a person in their prime catching a disease and then being gone within a matter of days, and what could be harder to comprehend is not knowing why a loved one would have an awful outcome while so many others recover.

Additionally, beyond the headline in the news article, there is no information on prevalence, nor relative risk, nor the fact that this is such a rarity that it does make the news. Let us dig into this phenomenon a little deeper.

THE RISK OF DYING FROM COVID-19

What went unreported was the overall distribution of deaths by age. To put this into some perspective, the CDC maintains a weekly provisional count and analysis at their website; the data in Figure 13 is current up to September 2, 2020.[79]

Displaying this data as a pie chart, you can clearly tell that over 3/4 of the deaths by COVID-19 in the U.S. occur with those over the age of 65 and that those under the age of 35 make up literally 1% of the deaths. Below 24 years of age, the figure states 0% rounded down; the exact calculation is 377 deaths under 24 years of age, out of 175,866 deaths, yielding a rate of 0.2%.

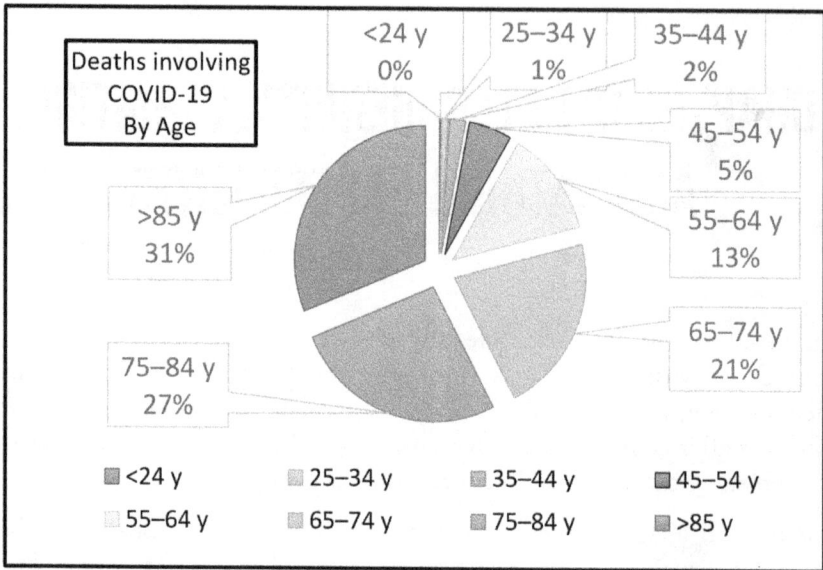

Deaths involving COVID-19 By Age

<24 y	25–34 y	35–44 y	45–54 y	55–64 y	65–74 y	75–84 y	>85 y
0%	1%	2%	5%	13%	21%	27%	31%

Legend: ■ <24 y ■ 25–34 y ■ 35–44 y ■ 45–54 y ■ 55–64 y ■ 65–74 y ■ 75–84 y ■ >85 y

Figure 13: Frequency of death by age range. Over 75% of deaths by COVID-19 in the United States come in patients over age 65, and patients under the age of 35 make up less than 1%. Percentages are rounded to the nearest round number [79].

The percentage of COVID-19 deaths in anyone under 55 is 8 percent. A Cambridge UK statistician named David Spiegelhalter has eloquently summarized the UK NHS Office of National Statistic's data on the risks of dying from COVID-19 among people who do currently have the disease. Overall, a positive diagnosis roughly doubles your risk of dying this year[80]. Put another way, getting COVID-19 is like packing a year's worth of risk into a week or two, and within any age group, the risk is disproportionately borne by the already chronically ill and those with pre-existing conditions.

UNPACKING THE MYSTERY OF HETEROGENEITY

The term "heterogeneity" essentially means diverse. There is a wide diversity of responses, from no symptoms at all to death within days of infection, and the differences between individuals fall in a few large categories. First would be existing pre-conditions; in the prior chapter, we looked at how the U.K. OpenSafely Project examined these pre-conditions in detail, such as individuals having an additional risk of serious disease and death due to BMI, diabetes, cardiovascular disease, and cancer.

The next section will cover a second factor; the possibility of some level of individual resistance from pre-exposure, presumably from other coronaviruses that cause the common cold. Afterward, there will be a detailed analysis of the individual correlates to mild disease. Lastly, a look at the underlying genetic contribution to severe disease, applying the genetic microarray technology introduced in Chapter 3.

INDIVIDUAL RESISTANCE FROM PRE-EXPOSURE

Several studies have reported strong T-cell responses to SARS-CoV-2 from individuals who have never been exposed to the virus. It is suspected that it was prior exposure to common cold coronaviruses generating CD4+ T-cells that provide some level of protection against SARS-CoV-2 infection or the severity of COVID-19. One study suggests there is pre-existing T-cell memory in 20% to 50% of individuals and concludes that "variegated T-cell memory to coronaviruses that cause the common cold may underlie at least some of the extensive heterogeneity observed in coronavirus disease 2019 (COVID-19)."[81]

In a different study using frozen samples taken from healthy individuals in 2018, researchers discovered cross-reactive T-cell activity to SARS-CoV-2 in 28% of the samples; they noted that since these were preserved samples, the 28% number could be considered a lower boundary.[82] And in a preprint (not peer-reviewed), CD4+ T-cells cross-reactive to SARS-CoV-2 were observed in 38% of the healthy samples tested.[83]

At this point, it is important to distinguish the difference between the presence of antibodies against SARS-CoV-2 versus specific memory T-cells that are ready to respond to an infection. Long-lasting immunity is conferred by the presence of the cross-reactive T-cells. To exactly what level of protection is unknown, but already identification of the specific protein sequences which the T-cells recognize is being teased out.[84]

ANALYZING THE GENOME FOR MUTATIONS

It is intriguing and important to consider that natural human genetic variability could influence the course of SARS-CoV-2 — in particular, the cases of severe COVID-19 in younger individuals.

In Chapter 3 we reviewed the human genome project, as well as the advent of microarray technology a few years prior to the invention of next-generation sequencing (NGS). The microarrays identify DNA variants (called single nucleotide polymorphisms or SNPs, pronounced "snips"). There are a few million "common" SNPs; places along the chromosome where 1% or more of people have slightly different DNA from what others have. Then there are over ten million other SNPs that are called "rare," which are based on the same somewhat arbitrary cut-off point.

Studies that use these microarray "SNP Chips" to survey hundreds of thousands or millions of SNPs of many people are known as GWASs or Genome-Wide Association Studies. Most of the time, investigators are looking for genetic telltale mutations that are related to disease — its frequency or severity among people with particular SNP patterns.

The first GWAS to successfully connect a disease to a SNP pattern took place in 2005. The eye disease Age-related Macular Degeneration (AMD) was already known to have a genetic component. By current standards, this was a tiny study: 146 individuals were genotyped for 116,000 SNPs. However, the study was big enough to succeed: A single rare SNP that affects a blood-clotting protein was shown to greatly increase the risk of developing AMD.[85]

While this research unleashed a torrent of GWAS studies, many proved to be disappointing — either finding nothing of interest or discovering possible links that turned out to be false leads. It is now recognized that genetic influences on disease are usually subtle and complex; this nuance in these interacting genetic factors thus requires very large GWASs, sometimes exceeding 1 million participants and the application of sophisticated data analytics and statistics. To date, geneticists have successfully identified SNPs associated with devastating diseases such as cancer, Alzheimer's Disease, schizophrenia, cardiovascular disease, and diabetes.

Next comes a genetic investigation of individuals (all males below the age of 35 in this case) who died or had severe cases of COVID-19 due to the same mutation in their genome.

TRAGIC MUTATIONS IN IMMUNE-RELATED GENES

The genetics of two families in the Netherlands, each with two brothers who suffered from severe COVID-19, have already been utilized to discover new aspects of how some individuals suffer fatally from this infection. The case illustrates the promise of studying the genomes of individuals and their families who have sudden and severe cases of COVID-19.

Two unrelated families in the Netherlands each had two relatively young males who contracted COVID-19, with one individual dying.[86] These men were all less than 35 years of age with no pre-existing conditions, but the course of COVID-19 was so severe that each one was admitted to the ICU. All required between one and two weeks of ventilation. Additional family members volunteered their genomic DNA in order to aid in the analysis of rapidly performed, whole-exome data (that is, the 50 million bases of DNA that encode the 20,000 proteins in the human genome).

The same rare mutation on the X-chromosome was identified across all four affected individuals. An immune-system gene called *TLR7* for Toll-Like Receptor 7 was mutated. Because males have only one X-chromosome (inherited from their mother), any mutation on an X-chromosome gene like TLR7 will be fully expressed. While the men did not have major health problems prior to 2020, infection by SARS-CoV-2 meant a serious case of life-threatening COVID-19 disease.

The researchers did further experiments by examining the effect of impaired *TLR7* function, pointing to the loss of a few important immune system regulators, as well as tying findings into previous research into SARS-CoV and MERS-CoV. These mutations are likely to be very rare in the population, and there is a hypothesis that the additional copy of *TLR7* across all females (a concept called gene dosage) could contribute to the lower risk of death from COVID-19 in females.

Another study looked at a set of 659 individuals with severe COVID-19 (ages ranging from one-month old to 99-years old, 75% men and 25% women; 91 of the 659 individuals died from COVID-19) and compared their genomes with a set of 534 individuals with mild or asymptomatic disease. In 23 of the 659 people with severe disease (3.5%), mutations were discovered in 13 genes involved with interferon-related protection. These mutations were never present in individuals with mild disease.[87]

While this genetic defect in interferon-related immunity explains only a small portion of severe cases, this genetic analysis led to another line of research by the same group of researchers. They examined the potential for these severe COVID-19 cases to have their interferon-related protection affected by antibodies against interferons — another example of the immune system gone haywire from Chapter 7.

Of 987 patients with life-threatening pneumonia, scientists discovered anti-interferon antibodies in over 10% of them, and none of these antibodies were present in the 663 patients with mild COVID-19. Additionally, they looked for the presence of these anti-interferon antibodies in 1,227 healthy individuals and found them in only 4 (about 0.3%). Interestingly, of the severe COVID-19 patients with these anti-interferon antibodies, over 94% were male.[88]

Ongoing research into protective genetic mutations will be covered next.

ONGOING GENETIC RESEARCH

There will be additional natural experiments like this one. Two efforts will be highlighted next using microarray GWAS technology: One comes from the direct-to-consumer company 23andMe, and the other from an international effort to look at severe COVID-19.

The call for volunteers to self-identify was clear: Anyone with an existing 23andMe test could let the company know whether they had contracted COVID-19 and the level of severity, and the company would use their existing datasets of the million genotypes per individual to see whether any of the SNPs would correlate with the risk of infection or risk of COVID-19 severity.

There is a history of genetics affecting a person's susceptibility to different infectious diseases — for example, the aforementioned *CCR5* mutation conferring protection from infection by HIV or a particular mutation in the *FUT2* gene (affecting the presence of molecules that line the stomach and digestive tract) that greatly reduces the risk of being infected with Norovirus, which somewhat infamously causes violent dyspepsia and diarrhea on-board cruise ships.

23andMe sought to enroll 10,000 participants who contracted COVID-19 and was able to publish some early results quickly: Individuals with type O blood are 9% to 18% less likely to test positive for the coronavirus, and this protective effect holds up when controlling for complicating conditions and age (two known risk factors), as well as BMI and other factors.[89]

A larger study was organized by an international consortium called the COVID-19 Host Genetics Initiative; this group of geneticists aims to discover the genetic factors that determine the likeliness of infection of COVID-19, as well as severity and medical outcomes. From this information, scientists can derive clues for potential therapy, identify people who are at unusually high or low risk of death, and contribute knowledge about COVID-19.

In publishing their first results, they illustrated that they had discovered three regions of significant association after analysis of a total of 1,980 patients from Italy and Spain, examining 8.5 million genetic variants (SNPs).[90] One genomic region on chromosome 9 pointed right to the ABO blood group genetic determinants, replicating what 23andMe had discovered only a few months before, not only regarding the protective nature of Type O blood (quantified here as an Odds Ratio or relative risk of 0.65, meaning that 45 percent were more likely to not be infected when all other variables are controlled for), but also a risk factor of Type A blood, with an Odds Ratio of 1.45. In other words, if you have Type A blood, there is a 45 percent increased chance of you becoming infected.

The other area the association study found was on chromosome 3, a collection of genes (specifically called *SLC6A20, LZTFL1, CCR9, FYCO1, CXCR6,* and *XCR1*). Of these genes, one called *SLC6A20* is of particular interest to scientists. This gene encodes a transporter protein that interacts with the main receptor molecule on the surface of the respiratory tract and other organ cells which the SARS-CoV-2 virus uses to gain entry into the body: the protein encoded by the *ACE2* gene.

These are associations giving clues into a further detailed investigation. Individuals with Type O blood are not immune from disease; they only have a lower relative risk compared to others. People with Type A blood will not automatically be infected upon exposure; they only have a higher relative risk. As a powerful tool for further study, a host of other scientists who is studying COVID-19 in detail will take a close look at *SLC6A20* and its role in disease progression.

COVID-19 has the terrible characteristic of a wide range of outcomes for individuals as a function of many underlying variables — from age, to sex, to different health conditions, to genetic makeup, to amount of viral dose — resulting in asymptomatic individuals who do not even know they have been infected, others with mild cases of flu-like symptoms, moderate cases of shortness of breath, serious cases of pneumonia or Acute Respiratory Distress Syndrome (ARDS), on to individuals who experience a cytokine storm and organ failure ending in death.

The genetics of individuals can and do play a role and offer clues into how new therapeutics can and will be developed. This effort will take many years if not decades to unravel, but we have already begun.

CHAPTER 10: A BRIEF HISTORY OF THE VACCINE AND ITS REMARKABLE POWER

Evolution favors the survival of the wisest.

—Jonas Salk

Life in China in the 4th century CE (common era) was difficult and only made more so by a plague that caused societal upheaval in decimating 30% of the population. Causing large bumps and making people look grotesque, these symptoms were accompanied by fever and often blindness. These telltale symptoms of smallpox were recorded in Chinese history. Smallpox spread from China to Korea and then to Japan in the 6th century, and then across the Arab world to Spain and northern Africa in the 7th century. During the Crusades in the 11th century, it spread across Europe, and in the 16th century — due to colonization and the slave trade — smallpox spread to Central and South America as well as the Caribbean. In the 17th century, it spread to North America, and in the 18th century, it spread to Australia.

Smallpox changed history. In North America, it has been estimated that the case fatality rate for smallpox among Native American populations was above 80%. In the 16th century, smallpox was the leading cause of death worldwide. In the 18th century, smallpox has been estimated to have killed approximately 800,000 individuals in Europe each year.

Over this three-thousand-year history, royalty was affected as well. A shortlist from 18th century Europe includes Queen Mary II of England (who died in 1694 at the age of 32); Holy Roman Emperor Joseph, the ruler of the Austrian Habsburg Monarchy (who died in 1711 at the age of 33); King Louis XV of France (who died in 1774 at the age of 64); and Tsar Peter II of Russia (who died in 1730 at the age of 15, on the eve of his wedding).

The earliest record of inoculation comes from China in 1549 with a process that was eventually called variolation. Scabs from smallpox sufferers were ground into a powder, then blown up the noses of healthy individuals. There was a risk in performing this procedure with a fatality rate of about 1%; however, after recovering from a mild case of the disease, most individuals were then immune to infection. About two hundred years afterward, this technique spread to Europe.

In 1796, the English physician Edward Jenner observed that dairymaids who contracted cowpox were then immune to the much more serious smallpox. He purposefully used infectious material from a cowpox pustule from an infected dairymaid to infect the arm of a nine-year-old boy named James Phipps, who was the son of Jenner's gardener. A few months later, after the boy recovered (which we now know was important, as this time gave the adaptive immune system time to develop immunity), Jenner purposefully exposed the boy to smallpox — and he was protected. In 1801, Jenner published his findings in a report titled "On the Origin of Vaccine Inoculation," and the practice soon became widespread. In 1813, in response to hucksters in the United States selling fraudulent vaccines, the U.S. Congress passed the Vaccine Act; the first U.S. law concerning protecting consumers from faulty pharmaceuticals.

FROM ONE TO MANY VACCINES

Smallpox was eliminated in North America in 1952, declared eradicated worldwide in 1979, and labeled as "the biggest achievement in international public health" as well as "one of the most important advances in modern medicine."[91]

Currently, the list of successful vaccines that have eradicated many other scourges is long; one reference lists 27 vaccines against many serious diseases, including the causative agent of typhoid fever, polio, rabies, and the bubonic plague.[92] Every year, the WHO estimates that two to three million lives are saved thanks to the availability of vaccination.

To move from the discovery of the causative agent (whether a virus or a bacteria) to the research required, the preclinical studies conducted in model organisms like mice and higher organisms such as primates, and on to phase I, II, and III randomized trials in humans, takes many years. Additionally, more time and resources must be spent building manufacturing capacity, producing the final vaccine, and then gaining final regulatory and reimbursement approval before distributing the final product. The typical timeframe is 14 years; given the outbreak of coronavirus in early 2020, that would make a vaccine against the current pandemic available in 2034.

The range of time it took for existing vaccines currently available is at its longest 28 years (for the nasal-spray FluMist) and at its shortest 4 years (for the mumps vaccine in the 1960s). Additionally, there are many important infectious diseases for which no vaccine has yet been able to be developed, including HIV, malaria, Respiratory Syncytial Virus, and Dengue.

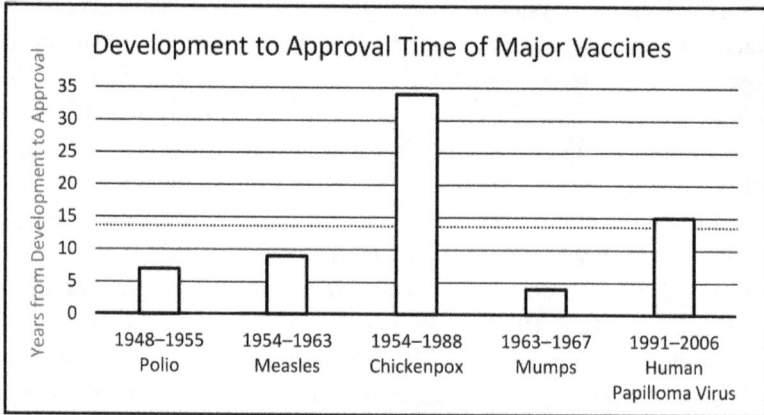

Figure 14: Development time of major vaccines. The shortest time to develop a vaccine was 4 years for mumps in the 1960s, and the average among this group of major vaccines is over 13 years.

A key advantage with the current pandemic is our history with SARS-CoV in 2003 and with the subsequent MERS-CoV in 2012. Basic and commercial research and development on both viruses, as well as potential vaccines, has given scientists a huge head-start. A long-lasting vaccine to prevent COVID-19 disease, or effective therapy to prevent death from COVID-19, is widely considered to be the final end to this pandemic.

Today, with the current vaccine effort against COVID-19, it is not known how long immunity will last, and those studies are built into the clinical trials that are ongoing. In its current thinking, the CDC's own data reports that no-one has been re-infected with COVID-19 within 3 months of initial infection.[93] The numbers of those with a confirmed diagnosis who have been re-infected are low; however, there are plenty of unknowns.

Next, we will discuss the methods of measuring vaccine effectiveness in animal models.

ANIMAL MODELS FOR DISEASE

The availability of ferrets for SARS and MERS is one key tool for evaluating the effectiveness of a potential vaccine, given that ferrets are susceptible to coronavirus infection. Research mice that have been genetically manipulated to express the human *ACE2* receptor (the mode of entry of SARS-CoV-2 into susceptible cells), which allows mice to be susceptible to coronavirus infection, are another key resource. Researchers can exploit these animal models to pursue vaccines that would produce a strong neutralizing antibody response to SARS-CoV-2 infection.

To date, while there has been active work on HIV, SARS-CoV, and MERS-CoV vaccines, none have made it through all of the clinical trial and efficacy hoops to achieve approval. With HIV, the virus itself attacks the immune system, which causes a major problem in getting the immune system itself to arm against it. For SARS-CoV, one challenge had been the outbreak started in December of 2002 and then ended by 2004. Unlike SARS-CoV-2, the original SARS caused severe illness rather than a range of symptoms, and there was no pre-symptomatic or asymptomatic spread from one person to another, and thus relatively limited in spread and vaccines could not be tested during active infection.

Several vaccines against SARS were developed, however, these projects ended up not being continued due to the simple fact that there was no method to show the vaccine's efficacy in human trials, given that no one was contracting the disease.

The story is similar for MERS. However, MERS saw periodic outbreaks in the Middle East after its initial spread in 2013, and a notable outbreak in Korea occurred in 2015. Each of these periodic outbreaks was relatively small in number (less than a few hundred), although the infection fatality rate for MERS is much higher than either SARS or SARS-2 (about 35% of those infected with MERS die of the disease). Again, with a limited number of infected humans, there was no clear pathway for a randomized, controlled trial to evaluate the efficacy of preventing infection among healthy individuals. Because of the low number of infections by MERS even among high-risk individuals (camel workers, families of camel workers, and hospital personnel), you would need over 100,000 participants for a random, controlled trial to be successful.

One notable success was for the Ebola virus; in late 2019, FDA approval was granted based upon a clinical trial conducted during a 2014 to 2016 outbreak in Guinea involving over 3,500 first- and second-level contacts of infected individuals. The vaccine was shown to be 100% effective in preventing transmission of the virus.[94] In that outbreak across West Africa, more than 28,000 were infected, and 11,000 died.

The Merck-produced vaccine ERVEBO® is based upon a live recombinant virus (called the Vesicular Stomatitis Virus, normally infecting horses and cattle) where an original viral coat protein is replaced with a protein from the Ebola virus.

FIVE MAIN METHODS TO PRODUCE A VACCINE

With tens of millions of cases of SARS-CoV-2 infection worldwide, over a million deaths, and ongoing infection, there are plenty of individuals in different countries worldwide to test the efficacy of a COVID-19 vaccine. The problem is time: How can all of the time-consuming steps (requiring 14 years) be accelerated in time to end this pandemic nightmare?

There are currently five different methods to produce a vaccine. The first is to use what is called an inactivated vaccine, which is the actual infective virus made non-infective through chemical or heat treatment. Producing a vaccine from live virus is difficult to accomplish reliably, and there are concerns about the process's safety with coronavirus since it has not been done before, although historically, this method has been used successfully for other viral and bacterial pathogens.

The second method is called live-attenuated vaccine, where the virus can still infect healthy cells. This is closest to a natural infection; you would literally create less-dangerous forms of the actual virus that causes disease. While it would create the longest-lasting immunity, there are clear safety concerns with taking this approach around coronaviruses in general — not only because it has never been done before for coronaviruses, but also because of the potential for the virus to revert back into a dangerous form.

The third option is using viral vectors or virus-like particles that are engineered viruses. These vectors are more like a single-use syringe, where the virus is only a delivery mechanism without any capability to reproduce more copies of itself.

Adenovirus has been studied in-depth since its discovery from adenoid tissue in 1953 and is commonly used for this purpose. Engineered adenovirus has a one-way payload and delivers a modified viral genome containing the code for producing the SARS-CoV-2 antigen in the cells it infects, producing an immune response. For the previously mentioned Ebola vaccine, using a cattle virus resulted in successful testing and approval. This approach, however, is not for use in immunocompromised individuals, as these viruses can induce immunity to the virus packaging itself. It is also more difficult to manufacture, as the engineered virus does not replicate and so literally billions of engineered viruses need to be administered to every individual patient in order to produce an immune response. It is worth noting that the current leading candidates for FDA approval are either implementing engineered adenovirus or the next method using RNA.

The fourth method is to use nucleic acid, whether RNA or DNA, packaged with nanoparticle or liposomal methods. Nanoparticles are what they sound like — manufactured materials that are minuscule compared to the size of a cell. These particles, literally tens or dozens of nanometers in size, interact with nucleic acid to protect it from degrading and causing cells to swallow it up in a process called endocytosis. Liposomes are chemically engineered fat molecules forming a water-in-oil bubble, also compatible with delivery to the cells the liposome encounters. This method is easy to design and produce and is also considered very safe. One drawback is the potential need for multiple doses to mount an effective immune response.

The fifth and last method is using protein subunits of the virus. These are relatively easy to design and produce and very safe to administer. However, their level of immune response tends to be relatively low compared to the others, requiring multiple doses or the use of an immune-boosting agent called an adjuvant.

The fact remains that the common cold, the first strain HCoV-NL64 (HCoV stands for human coronavirus), is estimated to have emerged around the year 1200. We have had four endemic coronaviruses with us (the other strains for the cold are called 229E, OC43, and HKU1), and they will be with us for the foreseeable future. No coronavirus to date has had a successful vaccine developed against it, although current manufacturers who have developed and tested vaccines against SARS-CoV-2 were working on MERS vaccines before the pandemic hit with varying levels of success (but these studies were laid aside, as mentioned previously, due to the lack of ongoing outbreaks).

For the four coronaviruses that cause the common cold, no vaccines have been successfully developed, partly because there are another 200 or so additional virus species and strains that cause cold symptoms, including the aforementioned adenovirus, rhinovirus, respiratory syncytial virus, and many others. Rhinovirus comprises about 75% of all colds, and there are some

150 strains in circulation. Developing vaccines against all these is clearly unfeasible.

Of the 32 vaccines in Phase I through Phase III clinical trials as of the Fall of 2020, none uses the first method of live inactivated virus, six use live attenuated virus (that is, the virus will replicate, but not dangerously), seven use adenovirus or another engineered virus as a vehicle, nine use RNA or DNA, and ten use protein.[77]

OPERATING AT THE SPEED OF A PANDEMIC

Against the backdrop of active vaccine development, the U.S. government declared its support for up to 14 manufacturers of vaccines with investment and purchase agreements, also enabling simultaneous manufacturing plant construction. Their ambitious goal: compress the time from the average 14 years down to 12 to 18 months.

Given that the shortest development time for any previous vaccine has been four years, even getting to the under-two-year mark would be considered very ambitious.

Yet by the Summer of 2020, a little more than six months on from the sequence of the virus in January, the first candidate vaccine from Moderna / NIAID, with the name mRNA-1273, entered into a Phase III trial. At that time, the researchers reported strong preliminary results from their Phase I trial with this vaccine.[95] The second candidate from Oxford / AstraZeneca named AZD1222 (formerly called ChAdOx1 nCoV-19 for Chimpanzee Adenovirus Oxford-1 novel coronavirus 2019) entered into the critical Phase III clinical trial in mid-August, with excellent results from their Phase I/II trials. Results from this latest trial are expected in October.[96] The third candidate close behind the others is BNT162 from BioNTech / Pfizer, which is currently in Phase III trials.[97]

Phase I typically involves 50 to 150 volunteers to assess safety, including side effects and other complications. At this stage, the ideal dosage is typically explored, with the amounts implemented in a relatively wide range (a 10-fold amount is not unusual). Phase II scales to a thousand or more volunteers and continues to consider safety and looks more closely at efficacy. (Efficacy is looked at during Phase I as well). Phase III is very large, on the order of 30,000 individuals, and the goal is to prevent moderate or severe COVID-19. In the jargon of a clinical trial, this is called the "primary endpoint" (what the top goal or goals are). Statistically, among 15,000 receiving a placebo and 15,000 receiving the vaccine, a significant reduction would be about 150 individuals in the placebo group getting moderate or severe COVID-19, with very few or no individuals in the vaccinated group being similarly affected.

For any given trial, there will be a flood of information for the FDA to analyze. In addition to the severity of COVID-19 disease among the two groups, they also must consider the number of asymptomatic cases, different levels of anti-SARS-CoV-2 antibodies, how many adverse events occur (everything from soreness around the injection site to headaches and fatigue), and how many deaths occur.[98] How vaccine response is measured will be reviewed next.

MEASURING RESPONSE TO A VACCINE

Antibody assays are used to determine how many neutralizing antibodies are present in vaccinated individuals and are widely available through commercial and academic clinical laboratories. Functional assays that determine if live SARS-CoV-2 coronavirus can be blocked from entry into cells are available to specialized research groups. These functional tests (as compared to diagnostic tests) utilize special cell lines (called Vero E6, which are from African Green Monkey kidneys); grown in laboratory tissue culture dishes, they are exposed to both the live SARS-CoV-2 virus and the serum from vaccinated volunteers. This functional work is a time-consuming, manual process, which requires special safety precautions since the live virus has to be maintained and measured, but it is very important to demonstrate active neutralizing activity.

A Phase III clinical trial is a large and expensive undertaking, whether taking place in the U.S. or in other countries. For a vaccine, 30,000 individuals are sought to enroll in these trials and there must be an active infection in the areas where individuals are recruited from in order to demonstrate the vaccine's protective effect. In the case of the Oxford / AstraZeneca vaccine candidate, researchers started clinical trials in South Africa and Brazil before beginning their trial in the United States, and these trials are ongoing.

Importantly, the measurement of success for FDA approval is a high bar: a statistically significant reduction of moderate or severe COVID-19 disease in the treated group compared to the control group. It is not just a matter of reducing infection with the SARS-CoV-2 virus tested with quantitative PCR.

When testing a cancer drug for late-stage disease effect, the size of a Phase III trial can be as small as a few hundred, depending on the prevalence of the disease. The parameters of what is to be measured (the aforementioned "primary endpoint") are established in advance.

For a preventative measure such as cancer screening (think of diagnostic tests such as a mammogram or a colonoscopy), the numbers are not in the several hundred; rather, they are in the tens of thousands. This is because, in a given group, only a tiny fraction of that group will be getting the disease (whether breast cancer or colon cancer), and you do not want the trial to drag on for many years.

For a vaccine Phase III clinical trial, the statisticians are presuming an infection rate over the time of the clinical trial period of a few months to be in the 1% range (thus, 150 out of 15,000 in the placebo group getting severe COVID-19 with none or a few getting it in the vaccinated group).

A LUCKY VOLUNTEER

In late August of 2020, a third vaccine entered into a second-half Phase II trial (called Phase IIb) with excellent Phase I/IIa results from Novavax / Emergent Technologies, called NVX-CoV2373. What is a little unusual about this vaccine is that I participated as one of 1,600 individuals selected for the Phase IIb trial.[99] This prospective randomized, controlled trial had five "trial arms" (five test conditions, only one of which is a placebo); thus, I have an 80% chance of receiving the active ingredient, which is, in this case, an engineered, full-length spike protein encapsulated in a proprietary nanoparticle (to aid both the stability of the protein and its immunogenicity) and simultaneously includes something called an adjuvant.

An adjuvant is a substance included in a vaccine to help boost an immune response. There are only a few FDA-approved ones, including ones based on an inactivated bacteria (called Bacillus Calmette-Guérin, which itself is under clinical investigation in over 20 clinical trials for its anti-COVID-19 properties) and water-oil mixtures. These adjuvants typically induce a strong antibody response called humoral immunity, but not a strong T-cell response called cellular immunity.

Novavax's vaccine includes an adjuvant they have derived from saponin, a plant-derived substance from the soapbark tree also found in most vegetables, particularly in legumes and quinoa. Saponin tastes bitter and has foaming properties; a soapbark extract called quillaia (with a high level of saponins) is used as an ingredient in root beer and crème soda (for flavor) as well as in carbonated sodas as a foaming agent.

Purified saponins have been shown to enhance T-cell-mediated immunity, which is of particular interest in the development of anti-cancer vaccines where cellular immunity to kill cancer cells is a high priority. In Novavax's publication of their Phase I study, a modified saponin adjuvant was demonstrated to have induced the production of neutralizing antibodies against the coronavirus spike protein in levels that were actually higher than the levels from sufferers of severe COVID-19 disease.[100] These levels were determined to be four times higher than the levels of antibody in typical convalescent sera. The Phase I trial also showed the presence of T-helper cells specific against the spike protein.

The Novavax vaccine candidate is unique in having a full-length spike protein, modified specifically against the mechanism it undergoes as a virus enters a cell. Through a detailed study of the mechanism of coronavirus HKU1's spike protein at the molecular 3D level, researchers learned that two things happen to the spike protein upon recognition by the *ACE2* receptor of a susceptible cell (whether in the lining of your respiratory tract or in your major organ systems in late-stage, severe COVID-19 cytokine storm). The first occurrence is the cutting of the protein as part of the normal viral entry by the *ACE2* receptor and a changing of the shape of the protein as it fuses to the cell membrane. Novavax takes advantage of these intimate behaviors of the spike protein itself to make very minor changes, to a total of six amino acids out of the total 1,272 amino acids that comprise the entire spike protein; four are made to prevent the cutting and two to lock the shape into a "prefusion" shape.

As mentioned previously, I had an 80% chance of receiving a vaccine; with it being a blind study, I should not be aware of whether or not I received a placebo (saltwater at physiological salt concentration and pH) or the vaccine. However, upon injection, there was a tingling sensation, along with some tenderness around the injection area in my upper arm for the following few days.

21 days later, I received a second dose. On this occasion, two of the trial arms were to receive either a high dose or a low dose of the vaccine; two of the other arms were to receive a placebo, meaning there was a 50% chance of getting a vaccine or not. The injection site was sore again, and the following day, I had some reaction to the vaccine: general fatigue, muscle soreness, and a headache. These symptoms decreased within the next two days, and the

clinical trial study organizer called me that first day to confirm my electronic reporting of these symptoms and their severity.

FIVE LEADING PHASE III CANDIDATE VACCINES

In the United States, the three current Phase III candidates are as follows: the Moderna / NIAID mRNA-1273, which uses RNA inside a proprietary nanoparticle agent; the BioNTech / Pfizer BNT162, which is similar to the Moderna RNA and nanoparticle approach; and the Oxford / AstraZeneca ChAdOx1/AZD1222, which uses a chimpanzee adenovirus.

Outside of the U.S., there are two major vaccines in Phase III trials in China. Sinopharm's vaccine is based upon virus inactivation and called CoronaVac, with their Phase III trials underway in Brazil and Indonesia. In July, CoronaVac was given the Chinese equivalent of an EUA, and at that time, it was reported that 40,000 ranking officials received the vaccine. The second Phase III vaccine from China, produced by CanSino Biologics, uses a modified adenovirus called Ad5 and is being developed in partnership with the Chinese Academy of Military Medical Sciences. Their Phase II results published in July show a strong immune response.[101]

In an unprecedented move the month before these results were published, the military announced this vaccine was approved as a "specially needed drug." Other officials and company employees are being offered the vaccine, and to what extent they are being informed of the risks involved is unclear. More recently, there has been news of an expanded scale of vaccination to several hundred thousand worldwide, even before Phase III trials have been completed.[102] A Phase III trial is currently underway in Saudi Arabia and Pakistan.

The last vaccine in Phase III trials is in Russia. The Gamaleya Research Institute (part of the Russian Ministry of Health) launched clinical trials in June of 2020 for a vaccine called Gam-Covid-Vac; even before a Phase III trial could begin, President Vladimir V. Putin announced in August that their health care regulator had approved the vaccine (renamed Sputnik V) to widespread scientific criticism; a few weeks later, it was clarified that the approval was conditioned upon a positive Phase III result. Publication of their Phase I data with a small number of volunteers showed an antibody response and mild side effects. A mixture of adenovirus Ad5 and Ad26, this vaccine technology was originally developed to combat MERS.

At this phase in testing vaccines, it is impossible to make any head-to-head comparison from early evaluations of neutralizing antibody activity, as each group of researchers uses their own independent methods and assays to evaluate both the presence of neutralizing antibody as well as measurements of T-helper cell activity. Nonetheless, with over 165 vaccines in various stages of pre-clinical (that is, tested in non-human primates such as macaque monkeys) to human randomized controlled trials (RCTs) and 32 vaccines in Phase I to Phase III trials in humans as of September of 2020, progress on what is expected to be a final solution to the coronavirus pandemic is on the horizon.

VACCINE CONCERNS

Vaccines have historically taken 14 years to develop. The estimated chance of success of any vaccine to be FDA approved is only 6% (or one out of 16).[103] In the case of Novavax, they have spent 30 years and over $1.5 billion to get to the Phase III trial stage for a flu vaccine and are still awaiting word of FDA approval. Even after a Phase III trial, the FDA's evaluation still means there is a significant chance of the drug or vaccine not being approved.

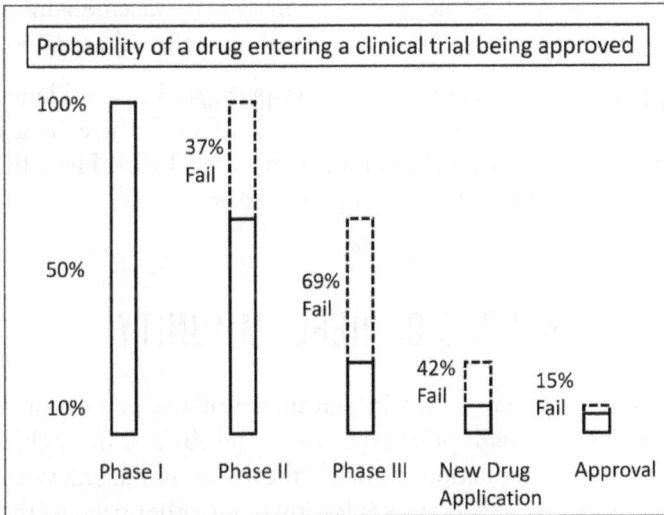

Figure 15: Chances of drug approval by stage. Less than 10% of all drugs entering a Phase I clinical trial get eventual FDA approval.

One concern from scientists is a side effect known as Antibody-Dependent Enhancement, or ADE. Antibody response to a virus like SARS-CoV-2 may bind very effectively to the viral target; however, instead of preventing entry into healthy cells, the antibody-coated virus acts as a facilitator of entry into the other white blood cells instead. Observed first for the mosquito-borne virus Dengue in the 1970s, antibodies against one strain of Dengue virus were shown to facilitate infection when added to the cell culture of other strains.

Thus, if you were vaccinated against Dengue Fever and later exposed to the real Dengue virus, you would actually get a worse case of Dengue Fever than if you had gotten no vaccine at all.

However, there is some debate as to whether ADE is only a laboratory-induced phenomenon for SARS-2 or a genuine risk; the argument can be made that while studies in primates demonstrate such an effect, primate and human immune systems have subtle and nuanced differences. Naturally, scientists are well aware of this effect and actively monitor the clinical trial results (specifically the type of T-cells involved) for evidence that this effect is occurring.

The rapid pace of progress gives reason for optimism. Nonetheless, with large Phase III trials and a high bar for the primary endpoint, the ultimate approval by regulatory bodies, the logistics of scaling manufacture, and the logistics of distribution, as well as the social dimensions of vaccine administration (including anti-vaccination and conspiracy theory efforts) are all to come.

Public support for vaccination has been dropping, as has confidence in both the scientific and government establishments. Could there be a different possibility to achieve the level of group protection (called Herd Immunity) without 60% of the population obtaining a vaccine?

A WORD ON HERD IMMUNITY

Herd immunity is where a certain percentage of the population has been immunized either through prior exposure to the virus or through a vaccine. When the basic reproduction number (the r0 or R-naught value) is less than 1, every infected person infects less than one other person. The upward exponential growth turns into exponential decline, and the viral spread ceases.

At a current estimated r0 value range of 1.4 to 4 (and an estimate of about 3), the number accepted among epidemiologists for herd immunity in a given population ranges from 50 to 75 percent (and based upon an r0 equal to 3 that number is 60 percent).

A growing body of research shows a significant number of people who test negative for neutralizing anti-SARS-CoV-2 antibodies have pre-existing T-cell immunity to the virus. Increasingly, it is suspected that pre-exposure to the coronaviruses that cause the common cold generate memory T-cells that are specific to SARS-CoV-2.

In the last chapter on heterogeneity, we touched on the finding that about a third of individuals never exposed to the novel coronavirus had some level of T-cell immunity. This has important implications for the percent of the population that needs vaccination for group immunity, currently understood to be 50 percent to 75 percent.

To take another example, SARS-CoV-2-specific memory T-cells were found in about half of the blood samples collected between 2015 and 2018, presumably from past coronavirus infections (i.e. the common cold)[104]. In yet another example in France, six of eight close family contacts of sick individuals did not develop antibodies against the virus; however, they did develop anti-SARS-CoV-2 memory T-cells.[105] A group in Sweden investigated moderate cases of COVID-19, finding both neutralizing antibodies and memory T-cells. However, researchers also looked at healthy individuals and asymptomatic family members of infected patients and compared their antibody responses to SARS-CoV-2 to their memory T-cells. Somewhat surprisingly, twice as many had memory T-cell responses compared to antibody ones.[106]

Why is this significant? In SARS, virus-specific memory T-cells persist for years, and these cells exhibit what is called stem-like properties (that is, properties analogous to stem cells, those cells that can differentiate into a panoply of other cell types with the right conditions).

Secondly, the previously mentioned serological surveys may well be undercounting — by a factor of perhaps 2-fold — the level of protection against SARS-CoV-2 infection in any given population tested only by antibody tests.

This may go a long way to explaining why so many individuals do not show symptoms, as their immune systems are already somewhat primed to defend against the new coronavirus invader. We may be closer to herd immunity than previously believed. Increasingly, there are researchers who believe that the generally accepted percentage of the population that needs to be vaccinated, mentioned earlier at 50 to 75 percent, may only be 25 or 40 percent instead.[107]

The race for a safe and effective vaccine is as urgent as ever and no less critical if the number of individuals to be vaccinated is 30% or 60%. With the potential availability of multiple vaccines to reach that threshold, in conjunction with approval of an effective monoclonal antibody treatment to prevent death from severe COVID-19 disease, the goal of returning to pre-pandemic life as usual may be closer than the end of 2021.

As we work our way there, the coronavirus foe is formidable: spreading throughout a hot and dry Summer as well as through a cold and humid Winter, causing a multi-system inflammatory syndrome in children and causing vasculitis with long-lasting consequences to organ health in others. We want to prevent all of this with a shot or two multiplied across hundreds of millions of individuals, so it's clear we have a long road ahead of us. So far the progress we have made is nothing short of remarkable.

CHAPTER 11: DEBATING THE CONTROVERSIAL

People like controversy because that's what sells.

— Miley Cyrus

Science is humble, but as with any workplace setting with smart and talented people, egos will get in the way. With a 24/7 endless news cycle, people being locked inside with constant streams of cable television, video chats, and social media, the same connectivity that scientists use to directly communicate breakthroughs with each other is also used to spread misinformation and fear. Fear is primal. The fear of death is the second most common fear people have (only beat out by the fear of public speaking).

In a place like the U.S. where personal liberty is held sacred, we have the perfect conditions for innovation. However, it is definitely not the best place for getting everyone to wear a mask and physically distance themselves. There is certainly room for scientists to be concerned about pandemic fatigue, as good news of vaccine development, better treatments, and scientific understanding are all coupled with a premature relaxing of the only interventions and preventatives we currently have: wearing a mask, keeping our distance, and avoiding enclosed and crowded spaces.

The list of controversies is long. Take mask-wearing as the first example: In an ill-advised move, the U.S. Surgeon General Jerome Adams said, in the months of February and March of 2020, that the public should not wear or buy masks to prevent the spread of the COVID-19 coronavirus.

From an interview on 2 March 2020:

> *"One of the things [the general public] shouldn't be doing is going out and buying masks...It has not been proven to be effective in preventing the spread of coronavirus amongst the general public...Folks who don't know*

119

*how to wear them properly tend to touch their faces a lot, and actually can
increase the spread of coronavirus. You can increase your risk of getting it by
wearing a mask if you are not a healthcare provider."[108]*

He followed the advice of the World Health Organization all the way
through the end of March.

As a scientist, I was highly skeptical of this original recommendation, as
it went counter to existing evidence. It was later proven to be incorrect.
Naturally, the emphasis then was to preserve medical-grade PPE for the
healthcare professionals, but to discourage face coverings of all kinds for the
general public, including homemade ones, was patently counterintuitive.

This stance was reversed:

*"…We now know from recent studies that a significant portion of
individuals with coronavirus lack symptoms. They're what we call
asymptomatic. And that even those who eventually become pre-symptomatic,
meaning that they will develop symptoms in the future, can transmit the
virus to others before they show symptoms. This means that the virus can
spread between people interacting in close proximity: for example, coughing,
speaking, or sneezing, even if those people were not exhibiting symptoms.*

*In light of this new evidence, CDC recommends and the [COVID-19] task
force recommends wearing cloth face coverings in public settings where other
social-distancing measures are difficult to maintain. These include places
like grocery stores and pharmacies. We especially recommend this in areas of
significant community-based transmission. It is critical."[109]*

Two large retrospective "what if?" scenarios are as follows. The first: if in
February, independent laboratories had been given the ability to develop
and validate their PCR tests based upon the CDC's or their design, to get
testing in-place and get ahead of the infection. The second: if officials had
recommended cloth face coverings in public settings in February instead of
in April. We will never know the result of these scenarios.

The Fall of 2020 is already shaping up to be one of confusion, with falling
test demand, debates over whether or not teachers are essential workers, and
stresses upon work-from-home parents everywhere; the list goes on. While
there is good evidence that this year's influenza season will be much more
mild thanks to physical distancing measures coupled with the widespread
use of masks, there will be some schools opening and businesses restarting
who will want tests that are fast, frequent, inexpensive, and easy to perform.

New tests and capabilities will come to the market and everyone will make adjustments.

Arguments will continue — for example around Hydroxychloroquine (the anti-malarial drug taken for decades by suffers from Lupus erythematosus) and azithromycin (a common antibiotic) with and without zinc for early-stage COVID-19 disease, or even its use as a prophylactic to prevent infection. As the previously referenced Marik protocol indicates, this is still a point of debate. Yet Dr. Malik took it off of his prophylactic protocol, as zinc takes many weeks before it appears in the body system-wide and simultaneous administration is unlikely to have much if any effect on existing results.

What is the proper perspective to maintain with the flood of information and misinformation swirling around? It comes down to a simple acknowledgment: At present, there are many things we do not know. For other topics, early data can be misleading. With still other remaining issues, the safest path is often the best one; with masks in particular, there is controversy on both sides with data to support both the effectiveness and ineffectiveness of mask-wearing.[110] But the cost-to-benefit ratio, where the cost is the inconvenience of mask-wearing and the benefit is that you have a lower risk of death...well, that's a price you should be willing to pay.

Next, we will discuss other areas where the middle ground perspective applies: the kinds of sampling that should be taken for diagnostics, whether surveillance testing is feasible, and how to look at re-opening schools for in-person learning.

THE MIDDLE GROUND OF SCIENCE

In COVID-19 diagnostics, the original sample type was a nasopharyngeal swab, where a long probe had to go all the way back through the nostril to touch the nasopharynx. There is debate and drive among laboratory scientists to use saliva as a sample type, to obviate the need for the sterile swabs, sterile tubes, and viral transport media (the liquid, sometimes pink and other times clear, which the swabs are stored in after collecting the sample). Would it be easy to measure whether saliva could be used for sample collection? One preprint paper states in its title that "Saliva is more sensitive for SARS-CoV-2 detection in COVID-19 patients than nasopharyngeal swabs."[111] And then there is another reference that states clearly its exact opposite point, that "Saliva is less sensitive than nasopharyngeal swabs for COVID-19 detection in the community setting."[112]

The truth is that it all depends; a meta-analysis combined a full 26 research papers and determined that saliva was less sensitive (91% sensitive with a 95% confidence interval between 80% and 99%) than nasopharyngeal swabs (98% sensitive with a 95% confidence interval between 89% and 100%)[113]. The researchers concluded that saliva tests "offer a promising alternative [to nasopharyngeal swabs, but] further diagnostic accuracy studies are needed to improve their specificity and sensitivity." The truth, as usual, is not clearly for or against, but somewhere in between.

One laboratory director also told me how impractical it is to work with saliva, given its viscosity and capability to handle small volumes of liquid. Viral transport medium is handled just like water (depending on additional additives sometimes it appears orange or red), and pipetting a tenth of a milliliter amount in the hundreds or thousands of samples is not an issue. However, the capability of 96- and 384-well liquid-handling robots to move saliva samples around would require time and effort to accommodate all of the potential problems with saliva being of varying thickness.

THE DRIVE FOR SURVEILLANCE

In advance of a commonly available vaccine, there is a drive to do mass surveillance testing in order to open up schools and businesses. One approach, described earlier, would involve only about 5 to 10 very automated, high throughput factories taking in 10's or 100's of thousands of samples per day and using the latest NGS instruments as the read-out instrument in huge 10-hour batch runs (instead of single batches of 96 or 384 samples at a time with existing qPCR equipment that take about an hour or two for each run).

Another method, sponsored by the Bill and Melinda Gates Foundation, is to work with an overnight shipping company, Amazon Corporation (with its immense customer logistics and order-taking capability), and a PCR technology commonly used for agricultural research. In crop development, companies such as Monsanto or Syngenta or Pioneer Hi-Bred will do hundreds of thousands of genetic analyses per day for seed improvement and on a massive industrial scale. Obtaining millions of genetic data points from hundreds of thousands of individual seeds is not uncommon. This technology uses a small, specialized plastic format called the TapeArray™ (trademark of LGC), and it could be repurposed for human use, specifically as a detection and diagnostic platform for COVID-19 PCR testing.

Or... it could not, as repurposing one for another isn't just a matter of plug-and-play. The TapeArray technology has never been used with human clinical samples before, and both the reagents and the instrumentation used for this application would need to go through all of the validation requirements previously discussed. The existing real-time PCR instruments used for COVID-19 testing have routine use for a variety of clinical testing applications, and under the FDA's EUA guidance documents, scientists would naturally prefer using clinical-grade instruments rather than others that are strictly used for research.

Then there are lateral flow immunoassays. Abbott, the Chicago-based diagnostic and therapeutic company, received an EUA for a new technology and a point-of-care reader that recognizes viral antigen via a paper-like strip. With the promise of a 15-minute, $5/sample test (a fraction of what PCR tests typically cost the laboratory performing the PCR test at $15-18 per sample, the remainder of the $100 reimbursement cost going to the laboratory for labor, other consumables, overhead, and profit), Abbott has the capability of producing some 50 million tests per month.

Additionally, per the EUA, this test is what is known as CLIA-waived, in that it is simple enough that a laboratory technician with a lower level of training can get consistent and reliable results with only the printed instructions. The EUA specifies that the test must be taken within 7 days of a patient showing symptoms and that it has a false-negative rate of 2.9% and a false-positive rate of 1.5%. Mass distribution of both the readers and the test strips may be much more feasible than getting samples to a few large centers, especially as the samples contain biohazardous material (i.e. potentially infectious samples).

But there is a larger problem here, and a significant one: What is the practical benefit of such a surveillance effort to justify the effort and expense to work our way through this pandemic? One report by the Rockefeller Institute and the Duke-Margolis Center for Health Policy came up with a number, reporting that it would take 193 million tests per month "to mitigate SARS-CoV-2 transmission risk to the point that the public could resume priority activities (e.g. in-person schooling) in a reasonably safe manner."[114]

The call is clear to ramp up surveillance, but there is the practical challenge of federal regulatory oversight: The FDA regulates medical devices (where diagnostics falls underneath), as well as therapeutics and vaccines, but the Centers for Disease Control has a budget and authorization for newborn screening efforts (for genetic disorders such as Cystic Fibrosis and Phenylketonuria), as well as for formulating the influenza vaccine every year.

And then underpinning them both is CMS, who will only pay for something, as mentioned before, if it is reasonable and necessary.

Is surveillance testing at scale reasonable and necessary? Not only would we have to have a prospective trial to test that hypothesis, but there would need to be the political will to pay for it as well.

Political debates will roil again and again, and we learn as we go along. One report analyzes an overnight camp in Georgia, where a full 75% (260 of 344 children and staff members) became infected in less than a week. At this camp, no cloth masks were used, nor windows and doors opened to increase ventilation in buildings, and the camp hosted a variety of indoor and outdoor activities including "daily vigorous singing and cheering." Within 5 days, the camp had to be shut down.[115]

Another contrasting report comes from a set of four camps in Maine, where 1,000 campers and staff members limited the spread to only 7 individuals. Rigorous testing before the camp began identified four individuals who were quarantined before arrival; campers quarantined in groups for the first two weeks upon arrival; and testing 5 days post-arrival identified three asymptomatic individuals (among both campers and staff) who were quarantined within their groups.[116] Information like this is valuable, informative, and actionable and tells us the importance of careful and routine testing, mask-wearing, and limiting the spread by group segregation.

Decisions on how a summer camp needs to be organized require discussion and debate in a healthy political environment. Scientific facts, however, are almost always tentative and subject to revision in the light of additional data. Mask wearing, maintaining distance, limiting indoor restaurant eating, and the closing of schools are several major actions where local governments have imposed various laws and regulations. The technical term is "Non-Pharmaceutical Interventions" (NPIs) and they have been studied in detail across countries to examine if any steps could be correlated back to infection and lethality rates for COVID-19 spread. So what does the data show?

A group of economists used something called reduced-form econometric methods, normally used to measure the effect of different policies upon economic growth across countries, to look at the effect of 1,700 local, regional, and national policies on the infection rate in six major countries (China, South Korea, Italy, Iran, France, and the United States).

Their work discovered that, if no policies were implemented, the growth of infected individuals would be some 38% per day![117] This shows a doubling rate of about 2 days.

The data shows some surprising results, the least of which is the overall effectiveness of these policies across countries. As of April 6, 2020, the U.S. had 365,000 confirmed cases, but due to N.P.I. policies put in place, an additional 4.8 million cases were prevented by these measures. In other countries, the numbers are similar: South Korea, 9.9K confirmed cases and 12 million fewer cases; China, 79.8K confirmed cases and 37 million fewer cases; Italy, 132K cases and 2.1 million fewer cases (all of these numbers are as of April 6, 2020).

In teasing out specific NPIs in this study, school closures did not have a significant impact, although the researchers were cautious to state that this factor needed further study because the data was not conclusive; banning large gatherings had a significant impact in South Korea and France.

Cumulatively, the total number by the researchers' model was in the range of 60 million infections being avoided. As we work our way out of this crisis, different political decisions have to be made in light of current knowledge, weighing the scientific data (tentative and subject to change with new data), economic considerations, and social impact.

IN-PERSON SCHOOLING — A HOT-BUTTON TOPIC

A hot-button topic is whether or not to open up schools and whether or not teachers are considered essential workers. While progress is being made on expanding testing, given its expense (the Medicare reimbursement to the testing laboratory is about $100 per test, while the cost of the reagents ranges from $15 to $40 per test) and FDA guidance on testing, currently the test is intended only for symptomatic individuals, not for everyone regardless of whether symptoms are present or not.

Our public health agencies have not had to face the twin challenges of a public health crisis and the need to regulate a generic surveillance test before. On top of these two issues come the questions of regulation and reimbursement. The Abbott BinaxNOW, $5, 15-minute test has recently been approved under EUA. Could this be put to use for surveillance, by getting the testing near to where the patients and the public are?

Given the simplicity of this approach, it could be further modified to include the pooling of samples and to include two or more sampling swabs per test strip, lowering per-sample costs further. As it is an antigen test (detecting viral protein rather than viral RNA), it is definitely not as sensitive, but fulfills the "Fast, Frequent, Cheap, and Easy" aims. Nonetheless, a less-sensitive test that can be broadly implemented at scale near the point-of-need could identify the individuals who need to be under quarantine.

It is clear that there are costs and benefits to surveillance. However, in the middle of a pandemic, there is no time to perform the necessary clinical trial for safety and efficacy of a diagnostic test used for a screening application. Please note that this is not the same as a test used to diagnose disease; this is a test meant to screen a population for a condition, where the consequences of accuracy (both false-positive and false-negative results) have a greater impact due to the sheer numbers of the individuals involved. On top of safety and efficacy come the economic evaluations of tests being reasonable and necessary, given that a public agency or other government organization would have to pay for the tests.

The American Pediatric Society, a historically conservative organization, weighed into the debate and came in strongly in favor of reopening, only to later modify their stance thusly (the text in bold per their website):

> With the above principles in mind, the AAP strongly advocates that all policy considerations for the coming school year should start with a goal of having students physically present in school. Unfortunately, in many parts of the United States, there is currently uncontrolled spread of SARS-CoV-2. Although the AAP strongly advocates for in-person learning for the coming school year, the current widespread circulation of the virus will not permit in-person learning to be safely accomplished in many jurisdictions.[118]

There is a long list of harms (if printed out, it runs to 23 pages in length), and there are so many specific recommendations and considerations for the wide range of pre-K through high school settings and circumstances. It is a complex balance, and until the pandemic passes, there are no easy answers.

CHAPTER 12: A FEW REASONS FOR OPTIMISM

We are making historic progress towards a vaccine...we have three vaccine candidates in that late stage, Phase III clinical trials, with tens of thousands of people getting enrolled. When that data comes in, depending on the rate of infection in the community...that will be reviewed by an independent board and then on to the FDA. These decisions will be driven by the standards of science and evidence, and the FDA's "gold standard".

— **Health and Human Services Secretary Alex Azar, September 2, 2020**

As of the Fall of 2020, we appear to be at the beginning of the end: We have several vaccines in full Phase III testing, faster and highly distributed diagnostics being authorized by a responsive FDA, and time-consuming and large clinical trials underway to test a multitude of therapeutic compounds and combinations of treatments. The advent of multi-treatment-arm trials from personalized, precision medicine has definitely changed the way vaccine development will take place from now on, and it is inevitable that the biology of this disease will be further unpacked, with new approaches that will have benefits for decades to come.

For example, it was structural biologists who tackled the difficult and fundamental problem of getting proteins along with additional sugar modification groups attached to them (called glycosylation) to crystallize. With accurate X-ray diffraction patterns, the 3-dimensional structure could be deduced, and this gave Novavax insight into which specific amino acids to change. Novavax's vaccine, currently called NVX-CoV2373, changed 0.5% of the amino acids in order to block the normal spike protein function of invading the susceptible cell's membrane wall. The result was a protein vaccine that the body could mount a long-lasting and highly effective immune response against, conferring protection with both antibody and cellular immune system responses.

In another example, the diagnostics industry responded to market signals without a top-down directive to develop at-scale responses for better, faster, and less expensive solutions to detect SARS-CoV-2. Manufacturing capacity constraints have been minimal, although periodic shortages of sampling swabs and plastic pipette tips (to name two examples) have occurred. Removing the swab requirement and using saliva (one Rutgers effort was issued an EUA along with the additional efforts described before) is one solution, along with new approaches such as the Abbott lateral flow immunoassay doing away with the need for pipette tips altogether, as well as the highly trained personnel needed to run PCR-based assays.

In the middle of this fight, there is a tough back-and-forth. Obtaining daily updates on the infection and death statistics, either at the world, country, state, or county level, is a common habit and behavior of everyone. News of different kinds, whether of new basic research or results from different clinical trials, is being commented upon and spun in different directions by different stakeholders. While the political environment and our current public healthcare status are volatile, we appear to be past the beginning and in a position to see the end.

Here are a few landmarks to keep an eye out for as indicators of progress.

WHAT TO LOOK FOR

There are several things to look out for as advances on different fronts take place. One is an FDA Emergency Use Authorization, which is a temporary, accelerated authorization to perform a test, use a therapy, or implement a vaccine only for the duration of this pandemic. The normal regulatory approval process takes 12 to 18 months, and there is an enormous amount of work being put in by the agency to prioritize and expedite the most promising approaches.

When considering the number of pivotal Phase III clinical trials for monoclonal antibody therapy and that several vaccine candidates are also steadily moving through Phase III trials, these both have the power to impact the speed at which discussions around opening up businesses and schools can begin. When new, effective therapies lower the number of fatalities and severe COVID-19 and the wider availability of vaccines lowers the incidence and severity of the spread, there will be a real and lasting improvement of our collective situation.

The second item to pay attention to is the study or data that underlies the key results. Thanks to the timing of the remarkable results obtained with dexamethasone (it occurred in late May, at the same time that the George Floyd protests began), the public did not hear that bit of good news. The reduction of death from COVID-19 with the use of dexamethasone by 30% among those already on a ventilator in the ICU and a 20% reduction from patients on oxygen is certainly newsworthy.

Given the flood of upcoming news, we also need to be on-guard against confirmation bias, which we will look at next.

GUARD AGAINST BIAS

Humans are experts at pattern matching, a function of basic survival embedded in our psyche. These internal biases are difficult to combat; we see patterns where none exist. Famously, we are terrible at assessing risk, and with fear being such a primal motivator, the click-bait titles and endless digital distractions hijack our higher-level thinking.

Consider what that first scientist, Ibn al-Haytham, said so long ago:

> *The duty of the man who investigates the writings of scientists, if learning the truth is his goal, is to make himself an enemy of all that he reads, and... attack it from every side. He should also suspect himself as he performs his critical examination of it, so that he may avoid falling into either prejudice or leniency.*

We all have our favorite ideas and concepts we may loathe giving up, but do not fall into either prejudice or leniency. I myself had high hopes for the cocktail of HIV antiviral medications, a combination of lopinavir and ritonavir. Both have been shown to be ineffective, although there are people still holding out for the benefit of hydroxychloroquine in the early stages of COVID-19 at a lower dosage. Yet, if consumer-grade disposable masks are only 70% efficient at reducing the spread of the coronavirus (compared to completely ineffective), I will wear a mask in public, as I may not know that I am infecting others until it is too late around two days later, should symptoms of infection occur.

Come up with critical examinations against your own beliefs, as al-Haytham stated. Take a realistic look at risk, a challenge due to our perception bias, and do your best to put that risk into perspective.

Can you disagree without being disagreeable? Can you take the opposite opinion of something you accepted as true when faced with additional data that opposes it? Do you have the attitude of a learner — that is, learning and growing, knowing the information may change with time?

TRUSTWORTHY ONLINE RESOURCES

Here are a few worthwhile resources and websites to access, in order of accessibility. The first group, science news, points to the news sections of primary literature in addition to specialty reporting. The second group, specialty editorial, gives specialty analyses of trends and news affecting their respective vertical markets. The third group, finding and using primary literature, is where the editorials are based and better than the editorial since it delves into the actual details.

For science news, *statnews.com* is a great life science and biotechnology news and editorial source; the Nature Research collection of journals has an excellent *News & Views section*.

Additionally, the journal *Science* is very useful.

A notable blog on the *Science Journal* website is Derek Lowe's excellent "In the Pipeline" blog about the drug discovery and pharma industry, with a great perspective on the progress being made in the therapeutics, vaccines, and diagnostics.

Editorial sites for diagnostics include *CAP Today* (the publication of an association, the College of American Pathologists, whose subtitle reads "Pathology–Laboratory Medicine–Laboratory Management"), *Diagnostics World News* (a publication of Cambridge Healthtech Institute, who organizes specialty conferences); and *GenomeWeb*, a life science industry publication, and its related publication specifically for diagnostics called *360Dx*.]

Editorial sites for therapeutics and vaccines include **Drug Discovery News** (a specialty print trade magazine) and **Fierce Family of Publications** (Fierce Pharma and **Fierce Biotechnology** in particular).

For the third group, finding primary literature to begin with requires knowledge of a few specialty websites and then often the journals with primary literature have a second barrier of access. During the COVID-19 pandemic, many have been made open access; however, you will still encounter publications of interest that may require you to ask help from a scientist-friend who has access, or to look at researchgate.net to see if anyone has posted it, or perhaps even to email the corresponding author for a copy (which they will often be glad to supply).

PubMed is a first stop along with a few preprint servers, *MedRxiv* and the related *BioRxiv*. One important caveat is that preprints will encourage rapid dissemination of information, but have not been through peer review, journal editing, and formal layout; thus, the quality can and does vary widely. In addition to the journals *Science* and *Nature*; the home pages of *The Journal of the American Medical Association* and *The New England Journal of Medicine* are worthwhile, as well as the British journal, *The Lancet*. One thing to consider would be a subscription to weekly email summaries of research reports from these journals; however, you must be warned that these publications publish a lot of material to be reviewed every week.

As a person who has read and studied primary literature for several decades, sometimes I only read a title; other times, I only read a title and abstract. Other times, it is the conclusions and a bit of the introduction I turn to. Of course, there are many times when I read an entire article and if needed, I will dig into the Supplementary Materials if they are available so that I can learn more about the data that did not make it into the paper, but which has valuable nuggets of information nonetheless. Like any other skill, it takes practice to get better and faster at comprehending scientific prose and methods of analysis, both technical and statistical. Yet you can be self-taught when exposed to relevant and timely information that can give the needed context beyond the news headlines.

Social media can be a boon or a bane, depending on who is in your news feed. There are notable voices on Twitter that cover policy (Scott Gottlieb @scottgottleibmd and Andy Slavitt @ASlavitt), the virologists / epidemiologists on the front-lines with a voice (Natalie Dean @nataliexdean and Trevor Bedford @trvrb), and physicians with a great perspective (Eric Topol @erictopol and Karol Sikora @profkarolsikora). Another great voice (one who is only on YouTube however) is John Campbell, Ph.D. (he has a doctorate in Nursing), who has a daily video breaking down research papers and country-specific trends.

These words have been written in the Fall of 2020; by the time you read this passage, the rate of infection may be similar to what it is now, it could be a small fraction, or gone altogether. Diagnostics may be widespread, breakthrough therapeutic options reducing the death rate and severity of COVID-19, and several vaccines being rolled out first to high-risk individuals and then the general population.

This coronavirus will be studied for decades to come, as will the long-term effects of COVID-19 in addition to all of the social impacts on many levels. The responses to this pandemic at all levels of governance will be analyzed and commented upon. Society will get back to whatever the new normal will look like, in fits and starts.

Whenever the next pandemic arrives, we will be ready.

AUTHOR BIOGRAPHY

With 20 years of life science industry experience across research and clinical markets, Dale Yuzuki is a well-known thought-leader in genomics and the author of *The Next Generation Technologist* blog. He has worked for several leading companies in genetics, including Illumina and Thermo Fisher Scientific.

Before his career in the life sciences, Dale taught high school biology and chemistry, as well as English as a Second Language courses overseas. In 2015, as a global scientific affairs specialist, he was the popular face of Thermo Fisher Scientific's *Behind the Bench* blog and the #LabChat video series; on YouTube, he's offered over 75 videos highlighting the groundbreaking research of leading scientists worldwide.

He has a Master's degree in education from UCLA and a second Master's degree in Cell and Molecular Biology from San Francisco State University, and lives in Greater Bethesda, Maryland with his spouse and three children.

For inquiries please email Dale directly at info@daleyuzuki.com. For an author interview about this book visit www.bit.ly/daleinterview and to signup for email updates visit www.bit.ly/signmeupdale

If you have enjoyed this book, a review on Amazon or your eBook provider of choice would be appreciated.

ENDNOTES

Chapter 1: The Humility of Science

[1] "Steve Jobs' dent in the universe—the shocking truth revealed!", SolveNext Blog, accessed October 20, 2020, https://solvenext.com/blog/steve-jobs-dent-in-the-universethe-shocking-truth-revealed
https://bit.ly/covidbook01

[2] Michael Shermer, "Rumsfeld's Wisdom," Scientific American, September 1, 2005, https://www.scientificamerican.com/article/rumsfelds-wisdom/
https://bit.ly/covidbook02

Chapter 2: The Biotechnology Revolution in Brief

[3] Watson JD, Crick FH. Molecular structure of nucleic acids; a structure for deoxyribose nucleic acid. Nature. 1953;171(4356):737-738. https://www.doi.org/10.1038/171737a0
https://bit.ly/covidbook03-1

Chapter 3: The Human Genome Project's Boost to Biotechnology

[4] Venter, C. and Cohen, D. (2004), The Century of Biology. New Perspectives Quarterly, 21: 73-77. https://www.doi.org/10.1111/j.1540-5842.2004.00701.x
https://bit.ly/covidbook04

[5] "Homo sapiens breast and ovarian cancer susceptibility (BRCA1) mRNA, complete cds", GenBank Entry, accessed October 20, 2020, https://www.ncbi.nlm.nih.gov/nucleotide/555931
https://bit.ly/covidbook05

[6] Stephens ZD, Lee SY, Faghri F, et al. Big Data: Astronomical or Genomical? PLoS Biol. 2015;13(7):e1002195. 2015 Jul 7. https://www.doi.org/10.1371/journal.pbio.1002195
https://bit.ly/covidbook06

[7] Lander ES, Linton LM, Birren B, et al. Initial sequencing and analysis of the human genome. Nature. 2001;409(6822):860-921. https://www.doi.org/10.1038/35057062
https://bit.ly/covidbook07

[8] Venter JC, Adams MD, Myers EW, et al. The sequence of the human genome. Science. 2001;291(5507):1304-1351. https://www.doi.org/10.1126/science.1058040
https://bit.ly/covidbook08

[9] Li H, Durbin R. Fast and accurate long-read alignment with Burrows-Wheeler transform. Bioinformatics. 2010;26(5):589-595. https://www.doi.org/10.1093/bioinformatics/btp698
https://bit.ly/covidbook09

[10] "3.8B Investment in Human Genome Project Drove 796B in Economic Impact Creating 310,000 Jobs and Launching the Genomic Revolution", Battelle Memorial Institute, May 10, 2011, https://www.battelle.org/newsroom/press-releases/press-releases-detail/3.8b-investment-in-human-genome-project-drove-796b-in-economic-impact-creating-310-000-jobs-and-launching-the-genomic-revolution
https://bit.ly/covidbook10

[11] Shelton JF, Shastri AJ, Ye C, et al. Trans-ethnic analysis reveals genetic and non-genetic associations with COVID-19 susceptibility and severity. medRxiv. Published online September 9, 2020:2020.09.04.20188318. https://www.doi.org/10.1101/2020.09.04.20188318
https://bit.ly/covidbook11

Chapter 4: NGS a Miracle of Technology

[12] Wetterstrand KA. DNA Sequencing Costs: Data from the NHGRI Genome Sequencing Program (GSP), accessed October 20, 2020, https://www.genome.gov/sequencingcostsdata
https://bit.ly/covidbook12

[13] "Novel 2019 coronavirus genome", Virological discussion forum, accessed October 20, 2020, https://virological.org/t/novel-2019-coronavirus-genome/319_
https://bit.ly/covidbook13

[14] Lu R, Zhao X, Li J, et al. Genomic characterisation and epidemiology of 2019 novel coronavirus: implications for virus origins and receptor binding. Lancet. 2020;395(10224):565-574. https://www.doi.org/10.1016/S0140-6736(20)30251-8
https://bit.ly/covidbook14

Chapter 5: Battling an Epidemic Through Molecular Diagnostic Testing

[15] Martin Andre Rosanoff, "Edison in his laboratory", Harper's Magazine, September 1932, https://harpers.org/archive/1932/09/edison-in-his-laboratory/
https://bit.ly/covidbook15

[16] Wiersinga WJ, Rhodes A, Cheng AC, Peacock SJ, Prescott HC. Pathophysiology, Transmission, Diagnosis, and Treatment of Coronavirus Disease 2019 (COVID-19): A Review. JAMA. 2020;324(8):782-793. https://www.doi.org/10.1001/jama.2020.12839
https://bit.ly/covidbook16

[17] CDC Public Health Image Library, Centers for Disease Control and Prevention, accessed October 20, 2020, https://phil.cdc.gov/Details.aspx?pid=23354
https://bit.ly/covidbook17

[18] Gniazdowski V, Morris P, Wohl S et al. Repeat COVID-19 molecular testing: correlation with recovery of infectious virus, molecular assay cycle thresholds, and analytical sensitivity. medRxiv 2020.08.05.20168963; https://www.doi.org/10.1101/2020.08.05.20168963
https://bit.ly/covidbook18

[19] Sethuraman N, Jeremiah SS, Ryo A. Interpreting Diagnostic Tests for SARS-CoV-2. JAMA. 2020;323(22):2249–2251. https://www.doi.org/10.1001/jama.2020.8259
https://bit.ly/covidbook19

[20] "The Market for Clinical LDT Services and LDT Supplies" market survey, Kalorama Information, September 24, 2020, https://kaloramainformation.com/product/the-market-for-clinical-ldt-services-and-ldt-supplies/
https://bit.ly/covidbook20

[21] Corman VM, Landt O, Kaiser M, et al. Detection of 2019 novel coronavirus (2019-nCoV) by real-time RT-PCR. Euro Surveill. 2020;25(3):2000045. https://www.doi.org/10.2807/1560-7917.ES.2020.25.3.2000045
https://bit.ly/covidbook21

[22] Robert Baird, "What Went Wrong with Coronavirus Testing in the U.S.", The New Yorker, March 16, 2020, https://www.newyorker.com/news/news-desk/what-went-wrong-with-coronavirus-testing-in-the-us
https://bit.ly/covidbook22

[23] "Emergency Use Authorization (EUA) information, and list of all current EUAs", Food and Drug Administration, October 20, 2020, https://www.fda.gov/emergency-preparedness-and-response/mcm-legal-regulatory-and-policy-framework/emergency-use-authorization
https://bit.ly/covidbook23-1

[24] "Pandemic and All-Hazards Preparedness Reauthorization Act of 2013 (PAHPRA)", Food and Drug Administration, November 8, 2018, https://www.fda.gov/emergency-preparedness-and-response/mcm-legal-regulatory-and-policy-framework/pandemic-and-all-hazards-preparedness-reauthorization-act-2013-pahpra
https://bit.ly/covidbook24

[25] "Infectious Diseases Society of America Guidelines on the Diagnosis of COVID-19", Infectious Diseases Society of America, May 6, 2020, https://www.idsociety.org/practice-guideline/covid-19-guideline-diagnostics/
https://bit.ly/covidbook25-1

[26] Steven Levy, "Bill Gates on Covid: Most US Tests Are 'Completely Garbage'", Wired Magazine, August 7, 2020, https://www.wired.com/story/bill-gates-on-covid-most-us-tests-are-completely-garbage/
https://bit.ly/covidbook26

[27] Wang W, Xu Y, Gao R, et al. Detection of SARS-CoV-2 in Different Types of Clinical Specimens [published online ahead of print, 2020 Mar 11]. JAMA. 2020;323(18):1843-1844. https://www.doi.org/10.1001/jama.2020.3786
https://bit.ly/covidbook27

[28] Kucirka LM, Lauer SA, Laeyendecker O, Boon D, Lessler J. Variation in False-Negative Rate of Reverse Transcriptase Polymerase Chain Reaction-Based SARS-CoV-2 Tests by Time Since Exposure. Ann Intern Med. 2020;173(4):262-267. https://www.doi.org/10.7326/M20-1495
https://bit.ly/covidbook28

Chapter 6: Transmission, Sensitivity, and Diagnostic Testing

[29] Park SY, Kim YM, Yi S, et al. Coronavirus Disease Outbreak in Call Center, South Korea. Emerg Infect Dis. 2020;26(8):1666-1670. https://www.doi.org/10.3201/eid2608.201274
https://bit.ly/covidbook29

[30] Lee S, Kim T, Lee E, et al. Clinical Course and Molecular Viral Shedding Among Asymptomatic and Symptomatic Patients With SARS-CoV-2 Infection in a Community Treatment Center in the Republic of Korea [published online ahead of print, 2020 Aug 6]. JAMA Intern Med. 2020;e203862. https://www.doi.org/10.1001/jamainternmed.2020.3862
https://bit.ly/covidbook30

[31] Zou L, Ruan F, Huang M, et al. SARS-CoV-2 Viral Load in Upper Respiratory Specimens of Infected Patients. N Engl J Med. 2020;382(12):1177-1179. https://www.doi.org/10.1056/NEJMc2001737
https://bit.ly/covidbook31

[32] Cereda D, Tirani M, Rovia F et al. The early phase of the COVID-19 outbreak in Lombardy, Italy Preprint arXiv.org Submitted 20 Mar 2020 https://arxiv.org/abs/2003.09320v1
https://bit.ly/covidbook32

[33] Richard, M., Kok, A., de Meulder, D. et al. SARS-CoV-2 is transmitted via contact and via the air between ferrets. Nat Commun 11, 3496 (2020). https://www.doi.org/10.1038/s41467-020-17367-2
https://bit.ly/covidbook33

[34] Lu J, Gu J, Li K, et al. COVID-19 Outbreak Associated with Air Conditioning in Restaurant, Guangzhou, China, 2020. Emerg Infect Dis. 2020;26(7):1628-1631. https://www.doi.org/10.3201/eid2607.200764
https://bit.ly/covidbook34

[35] Endo A, Centre for the Mathematical Modelling of Infectious Diseases COVID-19 Working Group, Abbott S et al. Estimating the overdispersion in COVID-19 transmission using outbreak sizes outside China. Wellcome Open Res 2020, 5:67 https://www.doi.org/10.12688/wellcomeopenres.15842.3
https://bit.ly/covidbook35

[36] Adam D, Wu P, Wong J et al. Clustering and superspreading potential of severe acute respiratory syndrome coronavirus 2 (SARS-CoV-2) infections in Hong Kong, 21 May 2020, PREPRINT (Version 1) available at Research Square https://www.doi.org/10.21203/rs.3.rs-29548/v1
https://bit.ly/covidbook36

[37] Lemieux J, Siddle KJ, Shaw BM, et al. Phylogenetic analysis of SARS-CoV-2 in the Boston area highlights the role of recurrent importation and superspreading events. Preprint. medRxiv. 2020;2020.08.23.20178236. Published 2020 Aug 25. https://www.doi.org/10.1101/2020.08.23.201782 36
https://bit.ly/covidbook37

[38] Zeynep Tufekci, "This Overlooked Variable Is the Key to the Pandemic", The Atlantic Magazine, September 30, 2020, https://www.theatlantic.com/health/archive/2020/09/k-overlooked-variable-driving-pandemic/616548/
https://bit.ly/covidbook38

[39] "Our Data", The COVID Tracking Project, accessed October 20, 2020, https://covidtracking.com/data
https://bit.ly/covidbook39

[40] Curt Devine and Drew Griffin, "Plenty of Covid-19 tests are available, but they're not being used", MSN powered by Microsoft News, August 26, 2020, https://www.msn.com/en-us/news/us/plenty-of-covid-19-tests-are-available-but-they-re-not-being-used/ar-BB18oKHF
https://bit.ly/covidbook40

[41] Tim Urban, "From 1,000,000 to Graham's Number", Wait But Why, November 20, 2014, https://waitbutwhy.com/2014/11/1000000-grahams-number.html
https://bit.ly/covidbook41

[42] Giamarellos-Bourboulis EJ, Netea MG, Rovina N, et al. Complex Immune Dysregulation in COVID-19 Patients with Severe Respiratory Failure. Cell Host Microbe. 2020;27(6):992-1000.e3. https://www.doi.org/10.1016/j.chom.2020.04.009
https://bit.ly/covidbook42

[43] "COVID-19 Support Group", Body Politic, accessed October 20, 2020, https://www.wearebodypolitic.com/covid19
https://bit.ly/covidbook43

[44] Tang Y, Liu J, Zhang D, Xu Z, Ji J, Wen C. Cytokine Storm in COVID-19: The Current Evidence and Treatment Strategies. Front Immunol. 2020;11:1708. Published 2020 Jul 10. https://www.doi.org/10.3389/fimmu.2020.01708
https://bit.ly/covidbook44

[45] Sinha P, Matthay MA, Calfee CS. Is a "Cytokine Storm" Relevant to COVID-19? JAMA Intern Med. 2020;180(9):1152–1154. https://www.doi.org/10.1001/jamainternmed.2020.3313
https://bit.ly/covidbook45

[46] "BIO COVID-19 Therapeutic Development Tracker", Biotechnology Industry Organization, accessed October 20, 2020, https://www.bio.org/policy/human-health/vaccines-biodefense/coronavirus/pipeline-tracker
https://bit.ly/covidbook46

[47] "COVID-19 Treatment Guidelines", National Institutes of Health, October 9, 2020, https://www.covid19treatmentguidelines.nih.gov/introduction/
https://bit.ly/covidbook47

[48] Heidi Ledford, "Coronavirus breakthrough: dexamethasone is first drug shown to save lives", Nature, June 16, 2020, https://www.nature.com/articles/d41586-020-01824-5
https://bit.ly/covidbook48

[49] "Convalescent Plasma", National Institutes of Health, October 9, 2020, https://www.covid19treatmentguidelines.nih.gov/immune-based-therapy/blood-derived-products/convalescent-plasma/
https://bit.ly/covidbook49

[50] "Passive Immunity Trial Of the Nation for COVID-19 (PassItOnII)", NIH Clinical Trials Database, August 27, 2020, https://clinicaltrials.gov/ct2/show/NCT04362176_
https://bit.ly/covidbook50

[51] Renn A, Fu Y, Hu X, Hall MD, Simeonov A. Fruitful Neutralizing Antibody Pipeline Brings Hope To Defeat SARS-Cov-2 [published online ahead of print, 2020 Jul 31]. Trends Pharmacol Sci. 2020;S0165-6147(20)30166-8. https://www.doi.org/10.1016/j.tips.2020.07.004
https://bit.ly/covidbook51

[52] Katie Thomas and Gina Kolata, "President Trump Received Experimental Antibody Treatment", New York Times, October 2, 2020, https://www.nytimes.com/2020/10/02/health/trump-antibody-treatment.html
https://bit.ly/covidbook52

[53] "Search for FDA Guidance Documents", Food and Drug Administration, accessed October 20, 2020, https://www.fda.gov/regulatory-information/search-fda-guidance-documents
https://bit.ly/covidbook53

[54] "Study Assessing the Efficacy and Safety of Anti-Spike SARS CoV-2 Monoclonal Antibodies for Prevention of SARS CoV-2 Infection Asymptomatic in Healthy Adults Who Are Household Contacts to an Individual With a Positive SARS-CoV-2 RT-PCR Assay", NIH Clinical Trials Database, October 9, 2020, https://www.clinicaltrials.gov/ct2/show/NCT04452318
https://bit.ly/covidbook54

[55] Boulware DR, Pullen MF, Bangdiwala AS, et al. A Randomized Trial of Hydroxychloroquine as Postexposure Prophylaxis for Covid-19. N Engl J Med. 2020;383(6):517-525. https://doi.org/10.1056/NEJMoa2016638
https://bit.ly/covidbook55-1

[56] Arshad S.and Zervos M. et al. Treatment with hydroxychloroquine, azithromycin, and combination in patients hospitalized with COVID-19 Int J Inf Dis 97, 396-403 (2020) https://www.doi.org/10.1016/j.ijid.2020.06.099
https://bit.ly/covidbook56-1

[57] "NIH halts clinical trial of hydroxychloroquine", National Institutes of Health, June 20, 2020, https://www.nih.gov/news-events/news-releases/nih-halts-clinical-trial-hydroxychloroquine
https://bit.ly/covidbook57

[58] "WHO discontinues hydroxychloroquine and lopinavir/ritonavir treatment arms for COVID-19", World Health Organization, July 4, 2020, https://www.who.int/news-room/detail/04-07-2020-who-discontinues-hydroxychloroquine-and-lopinavir-ritonavir-treatment-arms-for-covid-19_
https://bit.ly/covidbook58

[59] Catteau L, Dauby N, Montourcy M, et al. Low-dose hydroxychloroquine therapy and mortality in hospitalised patients with COVID-19: a nationwide observational study of 8075 participants. Int J Antimicrob Agents. 2020;56(4):106144. https://www.doi.org/10.1016/j.ijantimicag.2020.106144
https://bit.ly/covidbook59

[60] Cavalcanti AB, Zampieri FG, Rosa RG, et al. Hydroxychloroquine with or without Azithromycin in Mild-to-Moderate Covid-19. N Engl J Med. Published online July 23, 2020. https://www.doi.org/10.1056/NEJMoa2019014
https://bit.ly/covidbook60

[61] "Chloroquine or Hydroxychloroquine With or Without Azithromycin", National Institutes of Health COVID-19 Treatment Guidelines, October 9, 2020, https://www.covid19treatmentguidelines.nih.gov/antiviral-therapy/chloroquine-or-hydroxychloroquine-with-or-without-azithromycin/
https://bit.ly/covidbook61

[62] Gudiol C, Pujol M, Bandera A, Scudeller L, Paul M, Kaiser L, Leibovici L. Long-term consequences of COVID-19: research needs. Lancet Infect Dis. 2020 Oct;20(10):1115-1117. https://www.doi.org/ 10.1016/S1473-3099(20)30701-5
https://bit.ly/covidbook62

[63] Zhang P, Li J, Liu H, et al. Long-term bone and lung consequences associated with hospital-acquired severe acute respiratory syndrome: a 15-year follow-up from a prospective cohort study. Bone Res. 2020;8:8. Published 2020 Feb 14. https://www.doi.org/10.1038/s41413-020-0084-5
https://bit.ly/covidbook63-1

[64] del Rio C, Collins LF, Malani P. Long-term Health Consequences of COVID-19. JAMA. Published online October 05, 2020. https://www.doi.org/10.1001/jama.2020.19719
https://bit.ly/covidbook64

Chapter 8: Adjusting Standard of Care Treatment in Real Time

[65] Ben Elgin and John Tozzi, "Hospital Workers Make Masks From Office Supplies Amid U.S. Shortage", Bloomberg, March 17, 2020, https://www.bloomberg.com/news/articles/2020-03-18/hospital-makes-face-masks-covid-19-shields-from-office-supplies
https://bit.ly/covidbook65

[66] The Masks Now Coalition, accessed October 20, 2020, https://www.masksnow.org
https://bit.ly/covidbook66

[67] Alexandra Petri, "D.I.Y. Coronavirus Solutions Are Gaining Steam", New York Times, March 31, 2020, https://www.nytimes.com/2020/03/31/science/coronavirus-masks-equipment-crowdsource.html
https://bit.ly/covidbook67

[68] Johansson MA, Reich NG, Meyers LA, Lipsitch M. Preprints: An underutilized mechanism to accelerate outbreak science. PLOS Med. 2018;15: e1002549. https://www.doi.org/10.1371/journal.pmed.1002549
https://bit.ly/covidbook68

[69] Fraser N, Brierley L, Dey G, Polka JK, Pálfy M, Coates JA. Preprinting a pandemic: the role of preprints in the COVID-19 pandemic. bioRxiv. Published online May 23, 2020:2020.05.22.111294. https://www.doi.org/10.1101/2020.05.22.111294
https://bit.ly/covidbook69

[70] Richardson S, Hirsch JS, Narasimhan M, et al. Presenting Characteristics, Comorbidities, and Outcomes Among 5700 Patients Hospitalized With COVID-19 in the New York City Area. JAMA. 2020;323(20):2052-2059. https://www.doi.org/10.1001/jama.2020.6775
https://bit.ly/covidbook70

[71] Williamson, E.J., Walker, A.J., Bhaskaran, K. et al. Factors associated with COVID-19-related death using OpenSAFELY. Nature 584, 430–436 (2020). https://doi.org/10.1038/s41586-020-2521-4
https://bit.ly/covidbook71

[72] Covid Care Protocol, Eastern Virginia Medical School, accessed October 20, 2020, https://www.evms.edu/covidcare/
https://bit.ly/covidbook72-1

[73] Meltzer DO, Best TJ, Zhang H, Vokes T, Arora V, Solway J. Association of Vitamin D Status and Other Clinical Characteristics With COVID-19 Test Results. JAMA Netw Open. 2020;3(9):e2019722. Published 2020 Sep 1. https://www.doi.org/10.1001/jamanetworkopen.2020.19722
https://bit.ly/covidbook73

[74] Freedberg DE, Conigliaro J, Wang TC, et al. Famotidine Use Is Associated With Improved Clinical Outcomes in Hospitalized COVID-19 Patients: A Propensity Score Matched Retrospective Cohort Study [published online ahead of print, 2020 May 22]. Gastroenterology. 2020;S0016-5085(20)34706-5. https://www.doi.org/10.1053/j.gastro.2020.05.053
https://bit.ly/covidbook74

[75] "FAQ: COVID-19 and Ivermectin Intended for Animals", Food and Drug Administration, May 1, 2020, https://www.fda.gov/animal-veterinary/product-safety-information/faq-covid-19-and-ivermectin-intended-animals
https://bit.ly/covidbook75

[76] Beigel JH, Tomashek KM, Dodd LE, et al. Remdesivir for the Treatment of Covid-19 - Preliminary Report [published online ahead of print, 2020 May 22]. N Engl J Med. 2020;NEJMoa2007764. https://www.doi.org/10.1056/NEJMoa2007764
https://bit.ly/covidbook76

[77] Jeff Craven, "COVID-19 Therapeutics Tracker", Regulatory Affairs Professionals Society, October 16, 2020, https://www.raps.org/news-and-articles/news-articles/2020/3/covid-19-therapeutics-tracker
https://bit.ly/covidbook77-1

[78] Sanjay Gupta, "The mystery of why the coronavirus kills some young people", CNN, April 6, 2020, https://www.cnn.com/2020/04/05/health/young-people-dying-coronavirus-sanjay-gupta/index.html
https://bit.ly/covidbook78

[79] "Weekly Updates by Select Demographic and Geographic Characteristics", Centers for Disease Control and Prevention, October 1, 2020, https://www.cdc.gov/nchs/nvss/vsrr/covid_weekly/index.htm
https://bit.ly/covidbook79-1

[80] David Spiegelhalter, "How much 'normal' risk does Covid represent?", Medium, March 21, 2020, https://medium.com/wintoncentre/how-much-normal-risk-does-covid-represent-4539118e1196_
https://bit.ly/covidbook80

[81] Mateus J, Grifoni A, Tarke A, et al. Selective and cross-reactive SARS-CoV-2 T cell epitopes in unexposed humans. Science. 2020;370(6512):89. https://www.doi.org/10.1126/science.abd3871
https://bit.ly/covidbook81

[82] Sekine T, Perez-Potti A, Rivera-Ballesteros O, et al. Robust T Cell Immunity in Convalescent Individuals with Asymptomatic or Mild COVID-19. Cell. 2020;183(1):158-168.e14. https://www.doi.org/10.1016/j.cell.2020.08.017
https://bit.ly/covidbook82

[83] Braun J, Loyal L, Frentsch M, et al. Presence of SARS-CoV-2 reactive T cells in COVID-19 patients and healthy donors. medRxiv. Published online April 22, 2020:2020.04.17.20061440. https://www.doi.org/10.1101/2020.04.17.20061440
https://bit.ly/covidbook83

[84] Nelde A, Bilich T, Heitmann JS, et al. SARS-CoV-2-derived peptides define heterologous and COVID-19-induced T cell recognition. Nature Immunology. Published online September 30, 2020. https://www.doi.org/10.1038/s41590-020-00808-x
https://bit.ly/covidbook84

[85] Klein RJ, Zeiss C, Chew EY, et al. Complement factor H polymorphism in age-related macular degeneration. Science. 2005;308(5720):385-389. https://www.doi.org/10.1126/science.1109557
https://bit.ly/covidbook85

[86] van der Made CI, Simons A, Schuurs-Hoeijmakers J, et al. Presence of Genetic Variants Among Young Men With Severe COVID-19. JAMA. 2020;324(7):663–673. https://www.doi.org/10.1001/jama.2020.13719
https://bit.ly/covidbook86

[87] Zhang Q, Bastard P, Liu Z, et al. Inborn errors of type I IFN immunity in patients with life-threatening COVID-19 [published online ahead of print, 2020 Sep 24]. Science. 2020;eabd4570. https://www.doi.org/10.1126/science.abd4570
https://bit.ly/covidbook87

[88] Bastard P, Rosen LB, Zhang Q, et al. Auto-antibodies against type I IFNs in patients with life-threatening COVID-19 [published online ahead of print, 2020 Sep 24]. Science. 2020;eabd4585. https://www.doi.org/10.1126/science.abd4585
https://bit.ly/covidbook88

[89] "23andMe finds evidence that blood type plays a role in COVID-19", 23andMe Blog, June 8, 2020, https://blog.23andme.com/23andme-research/blood-type-and-covid-19/
https://bit.ly/covidbook89

[90] Ellinghaus D, Degenhardt F, Bujanda L, et al. Genomewide Association Study of Severe Covid-19 with Respiratory Failure [published online ahead of print, 2020 Jun 17]. N Engl J Med. 2020;NEJMoa2020283. https://www.doi.org/10.1056/NEJMoa2020283
https://bit.ly/covidbook90

Chapter 10: A Brief History of the Vaccine and its Remarkable Power

[91] "History of Smallpox", Centers for Disease Control and Prevention, accessed October 20, 2020, https://www.cdc.gov/smallpox/history/history.html
https://bit.ly/covidbook91

[92] Amanna IJ, Slifka MK. Successful Vaccines [published online ahead of print, 2018 Jul 26]. Curr Top Microbiol Immunol. 2018;10.1007/82_2018_102. https://www.doi.org/10.1007/82_2018_102
https://bit.ly/covidbook92

[93] "When to Quarantine", Centers for Disease Control and Prevention, September 10, 2020, https://www.cdc.gov/coronavirus/2019-ncov/if-you-are-sick/quarantine.html
https://bit.ly/covidbook93

[94] Henao-Restrepo AM, Camacho A, Longini IM, et al. Efficacy and effectiveness of an rVSV-vectored vaccine in preventing Ebola virus disease: final results from the Guinea ring vaccination, open-label, cluster-randomised trial. Lancet. 2017;389(10068):505-518. https://www.doi.org/10.1016/S0140-6736(16)32621-6
https://bit.ly/covidbook94

[95] Jackson LA, Anderson EJ, Rouphael NG, et al. An mRNA Vaccine against SARS-CoV-2 - Preliminary Report [published online ahead of print, 2020 Jul 14]. N Engl J Med. 2020;NEJMoa2022483. https://www.doi.org/10.1056/NEJMoa2022483
https://bit.ly/covidbook95

[96] Folegatti PM, Ewer KJ, Aley PK, et al. Safety and immunogenicity of the ChAdOx1 nCoV-19 vaccine against SARS-CoV-2: a preliminary report of a phase 1/2, single-blind, randomised controlled trial Lancet. 2020;396(10249):467-478. https://www.doi.org/10.1016/S0140-6736(20)31604-4
https://bit.ly/covidbook96

[97] Mulligan MJ, Lyke KE, Kitchin N, et al. Phase 1/2 study of COVID-19 RNA vaccine BNT162b1 in adults. Nature. 2020;10.1038/s41586-020-2639-4. https://www.doi.org/10.1038/s41586-020-2639-4
https://bit.ly/covidbook97

[98] Derek Lowe, "The Vaccine Protocols", In The Pipeline Blog, Science Translational Medicine, September 21, 2020, https://blogs.sciencemag.org/pipeline/archives/2020/09/21/the-vaccine-protocols
https://bit.ly/covidbook98

[99] "Evaluation of the Safety and Immunogenicity of a SARS-CoV-2 rS Nanoparticle Vaccine With/Without Matrix-M Adjuvant", NIH Clinical Trials Database, October 9, 2020, https://www.clinicaltrials.gov/ct2/show/NCT04368988
https://bit.ly/covidbook99

[100] Keech C, Albert G, Reed P, et al. First-in-Human Trial of a SARS CoV 2 Recombinant Spike Protein Nanoparticle Vaccine. medRxiv. Published online Aug 6, 2020:2020.08.05.20168435. https://www.doi.org/10.1101/2020.08.05.20168435
https://bit.ly/covidbook100

[101] Zhu FC, Guan XH, Li YH, et al. Immunogenicity and safety of a recombinant adenovirus type-5-vectored COVID-19 vaccine in healthy adults aged 18 years or older: a randomised, double-blind, placebo-controlled, phase 2 trial. Lancet. 2020;396(10249):479-488. https://www.doi.org/10.1016/S0140-6736(20)31605-6
https://bit.ly/covidbook101

[102] Peter Hessler, "Chinese Citizens Are Already Receiving a Coronavirus Vaccine", The New Yorker, September 29, 2020, https://www.newyorker.com/news/news-desk/the-november-surprise-of-chinas-coronavirus-vaccine
https://bit.ly/covidbook102

[103] Pronker ES, Weenen TC, Commandeur H, Claassen EH, Osterhaus AD. Risk in vaccine research and development quantified. PLoS One. 2013;8(3):e57755. https://www.doi.org/10.1371/journal.pone.0057755
https://bit.ly/covidbook103

[104] Grifoni A, Weiskopf D, Ramirez SI, et al. Targets of T Cell Responses to SARS-CoV-2 Coronavirus in Humans with COVID-19 Disease and Unexposed Individuals. Cell. 2020;181(7):1489-1501.e15. https://www.doi.org/10.1016/j.cell.2020.05.015
https://bit.ly/covidbook104

[105] Gallais F, Velay A, Wendling M-J, et al. Intrafamilial Exposure to SARS-CoV-2 Induces Cellular Immune Response without Seroconversion. medRxiv. Published online June 22, 2020:2020.06.21.20132449. https://www.doi.org/10.1101/2020.06.21.20132449
https://bit.ly/covidbook105

[106] Sekine T, Perez-Potti A, Rivera-Ballesteros O, et al. Robust T cell immunity in convalescent individuals with asymptomatic or mild COVID-19. bioRxiv. Published online June 29, 2020:2020.06.29.174888. https://www.doi.org/10.1101/2020.06.29.174888
https://bit.ly/covidbook106

[107] Apoorva Mandavilli, "What if 'Herd Immunity' Is Closer Than Scientists Thought?", The New York Times, August 17, 2020, https://www.nytimes.com/2020/08/17/health/coronavirus-herd-immunity.html
https://bit.ly/covidbook107

Chapter 11: Debating the Controversial

[108] US Surgeon General Jerome Adams video interview with Fox News hosts Steve Doocy and Ainsley Earhardt, "US Surgeon General urges Americans to stop buying, wearing masks amid coronavirus", Fox News, https://video.foxnews.com/v/6137596907001#sp=show-clips
https://bit.ly/covidbook108

[109] US Surgeon General Jerome Adams press conference, "User Clip: Surgeon General Adams' comments on face coverings", C-SPAN, https://www.c-span.org/video/?c4865964/user-clip-surgeon-general-adams-comments-face-coverings
https://bit.ly/covidbook109

[110] Hilda Bastian, "The Face Mask Debate Reveals a Scientific Double Standard", Wired Magazine, April 8, 2020, https://www.wired.com/story/the-face-mask-debate-reveals-a-scientific-double-standard/
https://bit.ly/covidbook110

[111] Wyllie AL, Fournier J, Casanovas-Massana A, et al. Saliva is more sensitive for SARS-CoV-2 detection in COVID-19 patients than nasopharyngeal swabs. medRxiv. Published online April 22, 2020:2020.04.16.20067835. https://www.doi.org/10.1101/2020.04.16.20067835
https://bit.ly/covidbook111

[112] Becker D, Sandoval E, Amin A, et al. Saliva is less sensitive than nasopharyngeal swabs for COVID-19 detection in the community setting. medRxiv. Published online January 1, 2020:2020.05.11.20092338. https://www.doi.org/10.1101/2020.05.11.20092338
https://bit.ly/covidbook112

[113] Czumbel LM, Kiss S, Farkas N, et al. Saliva as a Candidate for COVID-19 Diagnostic Testing: A Meta-Analysis. medRxiv. Published online May 27, 2020:2020.05.26.20112565. https://www.doi.org/10.1101/2020.05.26.20112565
https://bit.ly/covidbook113

[114] "A National Decision Point: Effective Testing and Screening for Covid-19", The Rockefeller Foundation, https://www.rockefellerfoundation.org/report/a-national-decision-point-effective-testing-and-screening-for-covid-19/
https://bit.ly/covidbook114

[115] Szablewski CM, Chang KT, Brown MM, et al. SARS-CoV-2 Transmission and Infection Among Attendees of an Overnight Camp — Georgia, June 2020. MMWR Morb Mortal Wkly Rep 2020;69:1023–1025. https://doi.org/10.15585/mmwr.mm6931e1
https://bit.ly/covidbook115

[116] Blaisdell LL, Cohn W, Pavell JR, Rubin DS, Vergales JE. Preventing and Mitigating SARS-CoV-2 Transmission — Four Overnight Camps, Maine, June–August 2020. MMWR Morb Mortal Wkly Rep. ePub: 26 August 2020. https://www.doi.org/10.15585/mmwr.mm6935e1
https://bit.ly/covidbook116

[117] Hsiang S, Allen D, Annan-Phan S, et al. The effect of large-scale anti-contagion policies on the COVID-19 pandemic Nature. 2020;584(7820):262-267. https://www.doi.org/10.1038/s41586-020-2404-8
https://bit.ly/covidbook117

[118] "COVID-19 Planning Considerations: Guidance for School Re-entry", American Academy of Pediatrics, accessed October 20, 2020, https://services.aap.org/en/pages/2019-novel-coronavirus-covid-19-infections/clinical-guidance/covid-19-planning-considerations-return-to-in-person-education-in-schools/
https://bit.ly/covidbook118